Topics In Just-In-Time Management

Marc J. Schniederjans
University of Nebraska-Lincoln

Allyn and Bacon

Boston • London • Toronto • Sydney • Tokyo • Singapore

Series Editor: Rich Wohl
Manufacturing Buyer: Louise Richardson
Production Editor: Cheryl Ten Eick

Copyright © 1993 by Allyn & Bacon
A Division of Simon & Schuster, Inc.
160 Gould Street
Needham Heights, Massachusetts 02194

Library of Congress Cataloging-in-Publication Data
Schniederjans, Marc J.
     Topics in just-in-time management / by Marc J. Schniederjans.
          p.   cm.
     Includes bibliographical references and index.
     ISBN 0-205-13600-1
     1. Production management. 2. Just-in-time systems. I. Title.
TS155.S3243 1992   92-10592
658.5'6—dc20   CIP

Printed in the United States of America

10   9   8   7   6   5   4   3   2   97   96   95   94   93

CONTENTS

PART III   A JIT Simulation Game   287

PREFACE

The subject of "just-in-time" (JIT) means many things to many people. Some feel it is an approach to production planning and control; to others it is a methodology to achieve manufacturing excellence; to still others it is a philosophy to guide everyday work activities; some businesses even view JIT as a winning strategy in the highly competitive marketplace of the 1990s. All businesses know that when a winning strategy comes along, out of competition, you use it to win, or have it used against you and lose. JIT is well recognized and respected by all as a winning strategy, philosophy, methodology, or approach. The lack of willingness (mostly because of lack of information on the subject) in the 1970s to use JIT has forced the U.S. to play catch-up in the 1980s. Much of the JIT implementation in the 1980s has only moved U.S. industries closer to a state of parity with the Japanese and other foreign competitors who use JIT. Much of this work centered on imitating what the Japanese do in Japan in hopes of making it work in the U.S. Obvious differences between Japanese and U.S. production philosophies, such as the U.S. orientation toward the use of computer technologies, have offered some challenges to those U.S. managers who have been trying to integrate JIT. Combining the 1980s experience with JIT in the U.S. and recent research on how to go about integrating JIT has now poised U.S. operations to extend JIT philosophy by making it a unique "American" philosophy.

Who Should Read This Book

This book is designed for faculty to use in teaching business and industrial management majors who are interested in learning how to use JIT and how to integrate JIT in modern business organizations for the 1990s. This book is designed as a JIT supplement for an upper level undergraduate or a graduate college course in production planning and control. It can also be used as a primary textbook for a topical course or a seminar in JIT management. The book is also of use to professional production and operations managers who want to learn about JIT and how it can be integrated with computer-based systems.

Organizational Structure Of This Book

The book is organized into three parts. The first part is introductory JIT material and requires no prerequisite knowledge. The second part of the book presents a series of advanced topics. While brief overviews of the advanced applications of JIT are presented in these chapters, students are expected to have completed at least a basic production/operations management course as prerequisite knowledge. The third part of the book presents an optional JIT in-class simulation game to help illustrate some of the basic JIT principles.

Acknowledgments

Regardless of who claims to be an author of a book, it is always a creative work of many individuals. I wish to acknowledge those people with whom I have had direct contact in the process of writing this book. I would like to express my gratitude to Barry Render of the Roy E. Crummer Graduate School of Business at Rollins College, who gave me the opportunity and trust to do this book. I would like to thank Richard J. Schonberger of Schonberger & Associates, Inc. (who introduced me to JIT) and Ellen J. Dumond of California State University-Fullerton, whose very helpful editorial comments helped to shape the final form of the book. I would also like to thank a number of operations practitioners, including Ken "Giant Man" Kahre, Edgar "Hercules" Wood, Daniel "Thor" Hillen, Alan "Hawkeye" McHugh, and especially Mark "Captain America" Helmer. While working at Sregneva Unlimited in St. Louis, these people altered my ironclad views on the applicability of JIT solutions to human resource problems. (Remember "Toad City"!) I would also like to recognize the significant contribution of my editor, Rich Wohl, who kept me on track and supported my efforts. I take full credit for all errors, omissions, and typos.

PART ONE

An Introduction to JIT

CHAPTER ONE

An Introduction to Just-In-Time

Chapter Outline

What is Just-In-Time? An Introduction

Why Study JIT?

How This Book is Designed to Aid in Understanding JIT

Who Should Read This Book

Organizational Structure of the Book

Organizational Structure of the Chapters

Why You Should Read This Book

Limitations of This Book

Learning Objectives

After completing this chapter, you should be able to:

1. Define what JIT is and how it differs from non-JIT operations.
2. Explain the key principles that make up the JIT philosophy.
3. Explain why JIT is a subject worth learning.
4. Describe the productivity cycling process and how JIT makes it work.
5. Define a number of Japanese terms used in JIT.

What is Just-In-Time? An Introduction

The subject of *just-in-time* (JIT) means many things to many people. Some business people feel it is an approach; to others it is a methodology; and to still others it is a philosophy, concept, or strategy. JIT is all these things and more. JIT's origins in Japan during the 1960s caused it to be initially considered as an approach to inventory management [1, p. 5]. (We will discuss JIT as an inventory management approach in Chapter 2.) Others viewed JIT as an approach to quality control. (We will discuss JIT as an approach to quality control in Chapter 4.) The nature of JIT, though, cannot be considered a limited inventory or quality control approach, but a multitude of approaches with applications throughout the business organization. JIT is not just for a couple of departments in an organization, it is for all departments in all types of organizations.

JIT's integrative nature started expanding from inventory and quality control into other areas of operations, like production scheduling. One commonly used JIT scheduling methodology is the *kanban* card system. (We will discuss JIT production scheduling systems in Chapter 3.) As industry and academic researchers explored the ramifications of JIT in the 1970s in Japan and during the 1980s in the United States and throughout the world, it became clear that the dynamic nature of JIT is not a limited methodology, but is made up in part of many methodologies. No longer just a Japanese method of inventory management, JIT is now an internationally known approach to excellence in production and operations management. JIT's pursuit of perfection can philosophically and strategically be seen to embrace all aspects of business operations, in all countries.

A simple definition for JIT is the successful completion of a product or service at each stage of production activity from vendor to customer just-in-time for its use and at a minimum cost. JIT can also be generally defined as a strategy or guiding philosophy whose goal it is to seek manufacturing excellence [2]. JIT is based on eight key principles:

1. Seek a produce-to-order production schedule.
2. Seek unitary production.
3. Seek to eliminate waste.
4. Seek continuous product flow improvement.
5. Seek product quality perfection.
6. Respect people.
7. Seek to eliminate contingencies.
8. Maintain long-term emphasis.

Let's examine each of these key principles. While introductory in nature, these principles provide a foundation on which each of the chapters of this book will build. The entire JIT philosophy is dynamically changing all the time and cannot adequately be expressed in a single chapter or even a single book. Also, please keep in mind that no single organization can embrace every aspect of JIT, but to a greater or lesser degree, all organizations can use these principles. They have application to both manufacturing and service organizations. They can be used to improve the production of custom products in **job shop** operations as well as homogeneous products in **repetitive manufacturing.** The extent to which an organization can embrace the JIT principles will help define the extent to which they can expect to share in its reported benefits.

Principle 1. Seek a Produce-to-Order Production Schedule In a *produce-to-order* system the manufacturer waits to produce products until the customer places the order. The goal of the system is to produce the finished goods just-in-time for consumption. Under this system, the item is immediately sent to the customer when completed. It doesn't become a carrying cost draining inventory items, but instead immediately generates sales from its waiting customer. The JIT operation produces only what is necessary in time for its use. In contrast, non-JIT operations using **produce-to-stock** systems rely on forecast projections of demand and place production into their inventory. Unfortunately, forecasts are often in error, and inventory that is held in stock by inaccurate forecasting is costly.

While all organizations can seek a produce-to-order system, they may never accomplish it perfectly. Lead times for customers waiting for an order may be quite prohibitive. Also, some organizations using produce-to-order systems can and do have buffer inventories of finished goods. This JIT principle, as well as the rest, are designed to motivate the organization to "seek" its unique level of implementation. JIT principles do not have to be completely implemented for an organization to be a JIT operation.

Principle 2. Seek Unitary Production Each unit of finished product is viewed as a separate order. The goal of the JIT system is a production lot size of one. Why have such small lot sizes? This principle allows production flexibility and reduces inventory costs. It is much easier to make minor adjustments in a unitary production system to meet shifts in demand, than to revise the non-JIT large-lot operation. Large-lot production requires more planning and lead time. Once a large lot size is set in the non-JIT system, an organization's production planning for equipment, human resources, and vendor contracts are usually also set. When a minor change is needed to adjust for a shift in demand, management in a non-JIT operation will be motivated to resist any minor deviation from the large lot production levels to avoid the cost and effort the changes might entail. The ability to make quick shifts to meet changes in demand avoids the costs of unwanted inventory during a decrease in demand and avoids the costs of missed sales during an increase in demand. We can view the unitary production as a means to avoid what the Japanese call *mura* or a production unevenness. By being able to make small unitary changes in production schedules, firms can avoid major shifts in inventory planning and human resource reallocations necessary in large lot production scheduling.

Another reason for seeking unitary production is to help reveal inventory and production problems. Raw materials, component parts, and subassemblies are all paced to arrive just-in-time for their use during **work-in-process (WIP)**. No buffer or excessive amounts of inventory are desired in the operation. If a defective component part is found or a worker accidentally wastes a part, they will have no buffer inventory to use in its place. The defective component will prevent a unit of finished product from being completed, and will therefore noticeably motivate the revealing of component problems. The sooner defective components can be revealed, the less rework will be incurred in replacing them, and the sooner the problem causing the defect can be eliminated.

Principle 3. Seek to Eliminate Waste Waste, referred to by the Japanese as *muda,* should be eliminated in every area of the operation. The goal is to use no more than the minimum amount of equipment, materials, and human resources necessary to achieve production objectives.

One of the factors that causes waste is imbalance or unevenness between actual and needed capacity. Many firms possess an excess (i.e., an imbalance) of equipment capacity as a preventive measure to cover for machines that have broken down. These excess machines are a waste of equipment and costly when not used. By instituting adequate preventive maintenance as suggested under JIT, machine down time might be eliminated, and the firm spared the need for the backup equipment. Extra workers to cover for absenteeism are a waste and should be eliminated. When properly motivated under JIT, teamwork effort will replace the need for extra workers. Defects requiring costly and time-consuming rework are a waste and should be eliminated. Most importantly, the imbalance between production and demand causes waste and should be balanced out. If we produce unwanted units (i.e., overproduce), we generate inventory that causes carrying costs, and wastes money. If we don't produce enough inventory, we cause stockout costs, and lose money, and possibly lose customers. All areas of waste in a JIT system are continually identified and eliminated.

Principle 4. Seek Continuous Product Flow Improvement Improving product flow is the same as improving productivity. In a JIT system the product flow goal is to eliminate bottlenecked processes and all problems that decrease the production flow. By eliminating idle time in production flow caused by imbalances in production activities or wasted effort by the workers, productivity increases.

Let's say that the time it takes a product to be completed (i.e., to flow through a facility) is a function of motion and work:

$$\text{Product Time} = \text{Motion Time} + \text{Work Time}$$

Those activities that take up *motion time* include handling materials, transporting goods, preparation time, waiting, and counting things. All of these activities are wasteful and should be eliminated. Only the activities of *work time* (e.g., fabrication, assembly, etc.) add value to the product and make it worth purchasing by the customer. In focusing effort on eliminating motion activities a JIT operation can improve product flow. How can JIT eliminate such basic material handling activities?   This can be accomplished through a continuous effort of self improvement in what the Japanese call the "5 S's" [3, p. 28]. The 5 S's are: proper arrangement (the Japanese term, *seiri*), orderliness *seiton,* cleanliness *seiso,* cleanup *seiketsu,* and discipline *shitsuke.* Continuously sorting through inventory and equipment to discard what is unnecessary (*seiri*) and arranging what is left in an efficient manner for use (*seiton*) will reduce future work motion activities. Similarly, an uncluttered work center (*seiso*) and well maintained equipment (*seiketsu*) will avoid effort to find and use work center facilities in a timely manner. Finally, and most importantly, these efforts for improvement must become habitual (*shitsuke*) for continued long-term improvement in performance.

Principle 5. Seek Product Quality Perfection The goal in a JIT system is the habitual seeking of zero defects. Under the JIT concept called **total quality control (TQC)** quality is an ongoing and never-ending pursuit of perfection in the product [4, pp. 47–82]. (We will discuss TQC in greater detail in Chapter 4.) In a JIT operation the monitoring of quality is chiefly left up to the workers who produce the product. The worker is the ideal member of the production team to see the impact of poor quality in materials, component parts, or subassemblies. The worker is also in an ideal position to observe and report problems concerning worker and machine-caused defects.

Embracing the TQC concept requires quality control activities of all members for the total organization. When defects are found, their causes are immediately determined and corrected. The **acceptance sampling** ideas of taking random samples of incoming goods and finished products to determine if an entire lot has obtained an acceptable quality level are dropped in a JIT operation. In its place TQC concept dictates a 100 percent inspection. Every piece of raw material, every component part, every subassembly, and every finished product must be inspected as it passes through the production system. In this way quality is built into the product rather than hoping it is screened out through random selection process.

Principle 6. Respect People People produce goods, systems do not. People are the most important asset a company has. In a JIT operation that importance is made clear to everyone. In a JIT system people are given greater responsibility to control work flow and greater authority to insist on product quality improvements than in non-JIT operations. In JIT operations workers are given the opportunity to control production with either stop/go switches at work centers or some type of management signaling system. In some operations where the stopping of an operation might cause excessive start up costs, worker centers have red (a serious problem) and yellow (a minor problem) lights that are used to signal management that a production or quality problem has been encountered and needs correcting. The Japanese term used to describe these lighting signaling systems is *andon.*

By placing these *andon* systems at the workers' command, management shows a commitment that quality is important and that the worker's opinion about quality is vital. The presence of the stop/go switches at work stations actually grants the worker the authority to force management to respond to production and quality problems that workers report. This is a substantial difference from the passive suggestion box approach of the past.

Principle 7. Seek to Eliminate Contingencies When management is concerned about having enough inventory to cover a surge in customer demand, they acquire contingency inventory that represents a waste if it is not used. When management is concerned about having enough workers to cover production requirements, they acquire a contingency supply of workers that represents a waste if not used. There are many of areas in an operation where management invests resources wastefully as a part of their contingency planning efforts. While it is certainly prudent to perform contingency planning, making wasteful investments in those contingency plans is avoided in a JIT system.

In a JIT system demand is a prerequisite for production scheduling. When a production schedule is established, management should plan to produce just what is stated in the schedule. Planning the use of inventory, equipment, and people is focused on the minimum allowable resource investment to achieve the desired production schedule. While this will place stress on the production system, the goal in the JIT operation is to use this stress in a positive effort to seek further improvements in the production system. Workers are more likely to suggest ideas for improving work flow when they are behind than when they are comfortably ahead of their production quotas. By eliminating contingencies like buffer inventory, the problems in work flow, scrap, and defective materials are revealed and corrected more quickly.

    Principle 8. Maintain a Long-Term Emphasis Since most of the other principles of JIT are continual, an investment in JIT should be viewed as a long term commitment. The benefits of a JIT system do not always reveal themselves in the short term. To implement the seven prior principles takes time. In the short term, some production costs may actually go up because of the learning that must take place. Changing worker habits, job responsibilities, job authority, and entire production systems can take a great deal of time. Some organizations have taken years to design and build plants around the JIT/TQC system. The new Saturn Corporation's facility in Spring Hill, Tennessee, is one example. This facility represents the largest single building in the world housing an entire automobile facility. It is completely based on JIT/TQC operation principles. This facility represents a long term commitment whose return will take years to retrieve.

    Because JIT's benefits can sometimes take a long period of time to be noticeable, managers must adopt tactics to continually motivate workers to think JIT. One tactic used to help workers and management observe the JIT benefits over the long term is an ongoing log. At the shop floor level of a JIT operation, it is common for department managers to maintain a log of past success with production and quality control projects. The log also contains the results of those projects and the benefits the department's efforts caused. Over the long term this log can be used to show new and old workers the commitment the department has and will continue to have to solve production problems and maintain product quality.

    Collectively, these eight key principles are only a brief introduction to JIT. Hopefully they illustrate how philosophically impossible but incredibly simple the many goals of JIT are in practice. Regardless of whether JIT is called a system, an approach, a concept, a strategy, or a philosophy, it is a way to help organizations achieve manufacturing excellence.

Why Study JIT?

In addition to JIT being a concept, method, approach, and philosophy, it is a strategy for success in business. Regardless of a manager's career objectives, the JIT strategy for success will benefit those who understand its use.

    One of the reasons why we need to study JIT is to re-learn how to successfully compete against international competitors, like the Japanese. The successful use of JIT is well documented in the literature [2, pp. 229-236; 5]. While a particular application might not prove JIT's universal success, the evidence of market dominance by the Japanese using JIT is all too clear to U.S. automobile manufacturers. The origin of JIT production techniques is often traced to the

Toyota Corporation of Japan [2, p.6]. As of 1990, the Japanese automobile manufacturers' market share of the U.S. automobile industry is equaling the market share of General Motors, the largest automobile manufacturer in the country. For the Japanese automobile manufacturers to accomplish this objective in a period of about twenty years demonstrates in part how powerfully successful JIT can be for organizations that embrace it.

The logic behind JIT market dominance strategy becomes evident as we examine the productivity cycling process in Figure 1-1. The application of JIT is

FIGURE 1-1 • *The Productivity Cycling Process*

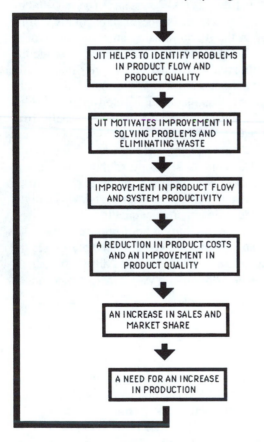

focused on identifying production flow and product quality problems. The identification of the problems in the JIT environment where contingencies are avoided forces management to quickly solve them. As production flow improves, productivity improves and costs decrease. As the quality of the products improves, the cost of rework, scrap, and refused orders by customers decreases. The decreased cost permits the organization to lower product prices. The lower product prices lead to an increased market share. An increase in market share forces a need for greater production. An increase in production can cause product flow and product quality problems. This then leads back to using the principles of JIT to further identify and solve the new problems caused by the increased production. Repeating the sequence of activities in Figure 1-1 results in a continued increase in market share. Since the JIT principles are meant to be continually applied, the logical result of the implementation of JIT is market dominance. Once market dominance is achieved, increased profit taking is made possible. Being patient and waiting for profit taking until market dominance is achieved in the longer term is one of the many lessons the foreign manufacturers have been teaching U.S. manufacturers. The basic profit motive that guides most U.S. organizations is very much a part of the JIT philosophy.

Another reason we need to study JIT is to learn how to avoid pitfalls that are unique to U.S. industry practices. In the U.S., where short-term profit taking is the norm, the effectiveness of the impact of implementing JIT can be diminished. As presented in Figure 1-2, Japanese organizations (which by culture are considered to be great savers) pass a larger share of the JIT cost reductions on to the customers through lower product prices than do similar U.S. organizations. This means the Japanese products can be priced lower than U.S. competitors even when both are using the same JIT strategy. As the proportion of the cost reduction benefits of JIT is reduced, so too is the opportunity to increase a firm's market share. Without the increase in market share the cycling process slows down and tends to stabilize improvement efforts and benefits. To maximize the productivity cycling process, producers must pass on to the customers the maximum amount of the cost reduction to achieve increased market share. Considering that 1990 labor costs in Japan were higher than in the U.S., and the advantage of reduced transportation costs in local markets, U.S. manufacturers using JIT currently have a unique advantage and opportunity of recapturing their market share lost to the Japanese. The key to recapturing markets is the willingness of U.S. organizations to invest (or pass on to the customer) the saved costs created by implementing JIT.

Wanting to be successful and avoiding pitfalls are only two reasons why we should study JIT. There are many other reasons including several additional positive benefits. JIT benefits organizations by reducing the costs of material, equipment, and people. At the same time, JIT improves product quality, productivity, and customer service. Many corporations have been benefiting from

FIGURE 1-2 • *Difference Between The U.S. and Japan in The Productivity Cycling Process*

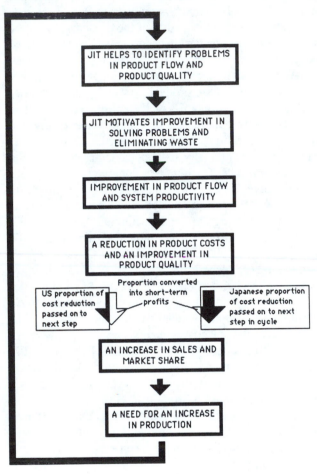

JIT for years. Some of the typical examples reported by U.S. organizations are presented in Table 1-1. Obtaining these and many other benefits requires the study of JIT.

TABLE 1-1 • *Some JIT Benefits Experienced by Corporations*

Corporation	JIT-Related Benefit
Hewlett-Packard, Sunnyvale, CA	Cut WIP requirements from three weeks to three days, and eliminated warehouse [2, p. 229].
General Electric, Louisville, KY	Cut lead time production for dishwasher products from six days to eighteen hours and reduced scrap and rework by 51 percent [2, p. 230].
Westinghouse, Fayetteville, NC	Inventories in motor-control centers cut from 4.2 months' supply to 0.89 month's supply [2, p. 232].
Baylock Manuf. Corp., Johnstown, CO	Installed computer-integrated JIT system and sales increased 400 percent because of improved scheduling [6, 98].
Tektronix Corp., Johnstown, CO	Reduced WIP by 34 percent, and reduced throughput time by 66 percent in only 75 days of the JIT program implementation [6, p. 102].
Computervision Corp., Manchester, NH	Reduced throughput time by 50 percent, improved quality by reducing waste by 30 percent, and reduced WIP by 66 percent [6, p. 111].
3M Corp., Weatherford, OK	Reduced unit cost of flexible disk media products by 30 percent and increased productivity by 60 percent [6, p. 112].
Ferro Manuf. Corp., Madison Heights, MI	Increased productivity by 46 percent, and reduced rework hours by 93 percent [6, p. 115].
Bytex Corp., Southborough, MA	Inventory cut by 43 percent and floor space reduced by more than 30 percent [7, pp. 51–53].

How This Book is Designed to Aid in Understanding JIT

Who Should Read This Book?

This book is designed for faculty to use in teaching business and industrial management majors who are interested in learning how to use JIT and how to integrate JIT in modern business organizations for the 1990s. This book is designed as a JIT supplement for an upper-level undergraduate or a graduate college course in production planning and control. It can also be used as a primary textbook for a topical course or a seminar in JIT management. Individual prerequisites vary by chapter. To maximize the educational benefits of this book, students are expected to have completed a basic introductory course in production/operations management. No prerequisite JIT knowledge is required. While manufacturing and service operation professionals might find this book topically interesting and quite useful, it is primarily written as a college textbook for use in the classroom. Its educational design focuses on general JIT learning objectives, terminology, and problem solving, rather than the more trade-oriented JIT books that focus on the practices and successes of individual companies.

Organizational Structure of This Book

The book is organized into three parts. The first part is introductory JIT material and requires no prerequisite knowledge. The second part of the book presents a series of advanced topics. While brief overviews of the advanced applications of JIT are presented in these chapters, students are expected to have completed at least a basic production/operations management course as prerequisite knowledge. The third part of the book presents an optional JIT in-class simulation game to help illustrate some of the basic JIT principles.

Part I, "An Introduction to Just-In-Time," consists of four chapters whose purpose is to introduce basic JIT concepts and methods. It seeks to spell out what JIT is and how it can be implemented. The purpose of Part I is to lay a basic background foundation on what JIT involves. Part II, "Integrating Just-In-Time," builds on the foundation presented in Part I. Specifically, Part II consists of three chapters whose purpose it is to explain how JIT can be integrated with more modern systems and technologies of the 1990s. As JIT's role expands to incorporate more functions in an organization, new areas of application and integration must be explored. Many of these new areas are explored by people whose research is not presented in textbooks for years after their application. Four of the more recent or currently popular areas of JIT application are presented in the chapters in Part II. Each chapter in this part of the book examines how JIT can be integrated into new areas of operations or integrated with new technologies. Part III, "A Just-In-Time Simulation Game," represents an optional set of exercises that requires a team of participants to mimic JIT principles in practice. The JIT simulation game is an in-class activity that will provide users with "hands-on" JIT experience. The game offers an opportunity to see how JIT principles result in a superior manufacturing philosophy when compared with other traditional philosophies.

The chapters in Part I are highly integrated. These chapters should be read in their entirety and in the order of their presentation. The design of this book permits flexibility in the reading of the chapters in Part II. Since the chapters in Part II are independent of one another, they can be read in any order, and any can be omitted. This permits the instructor or students an opportunity to focus only on selected topics. Part III is completely optional.

Organizational Structure of the Chapters

This book is designed for use by instructors who teach JIT. It is designed as a tool to provide support to the instructor's educational role and, therefore, to help make it easier for students to learn the subject. Each of the chapters has several sections in addition to its basic content:

Chapter Outline: an overview of chapter headings as a guide to content.

Learning Objectives: a listing of specific learning objectives that students should be able to accomplish after they have read the chapter. Students should read these before and after they have completed the chapter to ensure they have covered these major content areas.

Introduction: a section at the beginning of each chapter devoted to introducing the topic of the chapter and stating its purpose.

Summary: a summary section at the end of the chapter designed to review major content areas.

Important Terminology: a listing of terms or phrases that students should be familiar with in order to better understand the material in that chapter. The positioning of this chapter's related terminology is designed to overcome the major objection students have with a book whose advanced subject makes reading (i.e., constantly looking up prerequisite definitions) difficult.

Discussion Questions: a listing of questions designed to test conceptual learning. Instructors can use these for classroom discussion or assign them for homework. Students should use them as an alternative means of reviewing the chapter's conceptual material.

Problems: a listing of problems designed to test methodology and technical learning. Instructors can use these for homework assignments. Students should use them as an alternate means of reviewing the chapter's computational or logic material.

References: a listing of bracketed references used throughout the chapter. Most of the material in this book is drawn from research or books published in just the last few years. As such, the references will probably be available in university libraries. Students are encouraged to review this material to obtain additional JIT content.

Cases: one or two cases are presented as problematic situations related to the chapter's content. Instructors can use these cases as homework assignments. Students are encouraged to read these cases. They present a realistic problem situation that can help students see a more practical application of the chapter's content and aid in understanding how JIT can be implemented.

Why You Should Read This Book

All businesses know that when a winning strategy comes along, competition requires they use it, or have it used against them. JIT is well recognized and respected by all as a winning strategy. The lack of willingness in the 1970s to use JIT has forced the U.S. to play catch-up in the 1980s. Much of the JIT implementation work in the 1980s has only moved U.S. industries closer to a state of parity with the Japanese and other foreign manufacturers. Much of this work centered on imitating what the Japanese do in Japan in hopes of making it work in the U.S. Obvious differences between the Japanese and the U.S., and the U.S. focus on computer system technologies have offered some challenges to those U.S. managers who have been using JIT.

One of the differences between the Japanese use of JIT and the U.S. use has been U.S. management's reluctance to use JIT concepts and methods in service industries and in administrative functions. Most research and books on JIT focus on JIT as a concept for manufacturing and leave its implementation to service and the administrative areas up to the user's imagination. The more integrated JIT is in an organization, the more benefits can be derived from it. This book focuses separate chapters on each of these very important areas of application.

Another difference between the Japanese and the U.S. is U.S. manufacturing industries' greater orientation toward using computer-integrated systems. JIT doesn't require a computer system to operate, yet most U.S. manufacturers have been using computer-based systems in planning for decades. While software capable of supporting JIT was developed in the mid-1980s, little has been published on how to implement it outside of research articles. This book focuses content in several chapters on how JIT can be integrated with popular computer-based systems.

These topical differences have been selected as a focus for the subject of this book. That is why this book is called *Topics in Just-In-Time Management*. These differences have helped to slow the progress of JIT in the U.S. and currently offer major challenges for those who hope to take maximum advantage of what JIT offers an operation. Overcoming these differences and understanding how the integration of JIT with the more popular U.S. computer-based systems is possible are primary reasons for someone reading this book.

Limitations of This Book

No book in the market today, including this one, can claim to be the definitive one on the subject of JIT. JIT is changing too fast to be accurately captured in any book. This book attempts to lay a basic foundation of the principles, concepts, and methods of JIT in Part I. While this foundation may be incomplete for some, it summarizes most of the basics found in other JIT books on the market. The topics in Part II are meant to present a current collection of integration and implementation issues. It is by no means a complete collection of JIT topics, but seeks to focus on issues raised in recent research publications. It is hoped that by focusing on current JIT topics appearing in recent research, creative ideas and problem-solving suggestions presented in this book will quickly find application in industry.

Summary

This chapter presented an introduction to the key principles that make up JIT. This chapter also briefly explained why JIT is an important subject to study. Finally, the organizational design of the book, its chapters, and educational support structure were described.

Throughout this book we will refer to JIT as a combination of approaches, concepts, philosophies, strategies, or methodologies. Regardless of what we call JIT, we have seen in this chapter that it will require changes in all areas of a business operation. JIT requires a production schedule set to meet and not beat demand (Principle 1: Seek a produce-to-order production schedule) and a production effort that views each unit as a separate order (Principle 2: Seek unitary production). JIT seeks on one hand to improve product quality (Principle 5: Seek product quality perfection), while reducing production effort that is not essential for the completion of the product (Principle 3: Seek to eliminate waste). JIT also requires that the people who work together to produce a product respect one another as team players seeking a common goal (Principle 6: Respect people), while at the same time not permitting the players to err in their job performance (Principle 7: Seek to eliminate contingencies) and welcoming problems when workers find them (Principle 4: Seek continuous product flow improvement). JIT also requires users to have the patience to wait for the principles of the system to take hold and become a habit (Principle 8: Maintain a long-term emphasis). Resistance to Principle 8 by the short-term, quick-profit-oriented U.S. industries is one of the factors that caused JIT to take so long to be integrated into U.S. business operations. A good decade of experience with JIT has now poised the U.S. to extend JIT philosophy into making it a unique "American" philosophy. As we are often reminded by experts in motivation, the last four letters in American are "I can."

One of the major points JIT experts always try to emphasize is the importance of people, as briefly stated in Principle 6. People make decisions, people improve productivity, and people determine product quality. People are reading this book right now and deciding whether to read Chapter 2. If you don't read Chapter 2, you will have wasted effort in reading Chapter 1, and remember, JIT seeks to avoid waste. So read Chapter 2 and find out how JIT can be viewed as an inventory system whose philosophy has changed the world and permanently altered the subject of inventory management.

Discussion Questions

1. How does maximizing *seiri* minimize *muda*?

2. How does minimizing *mura* maximize productivity?

3. Is JIT a concept, a strategy, a philosophy, or a methodology? Explain.

4. How do produce-to-order systems avoid waste?

5. How does improved quality result in reduced production costs?

6. If it is physically impossible for a firm to make a 100 percent inspection of finished goods, can they still consider themselves a JIT operation if they embrace other JIT principles?

7. Why is JIT considered a dynamic process? Explain.

8. What is the advantage of unitary production as it is sought in a JIT system?

9. What are contingencies? How are contingencies wasteful? Where would they not be wasteful?

10. What is meant when we say "respect people" in JIT? Explain.

11. What is the productivity cycling process? How is it related to JIT?

12. What is one possible difference in the way U.S. manufacturers will implement JIT as opposed to the way Japanese manufacturers will implement it? Explain as it relates to the productivity cycling process.

Important Terminology

Acceptance sampling statistical sampling procedures used to determine if incoming lots of goods or outgoing lots of goods meet a predefined level of quality.

Andon a Japanese term for an alarm system used by workers to signal that a production line should be stopped by management to investigate a problem.

Job shop a type of manufacturing organization in which productive resources are organized according to function (e.g., painting department, assembly department, etc.) to produce a low volume or custom product.

Muda a Japanese term used to represent waste in a JIT operation.

Mura a Japanese term used to represent unevenness in an operation.

Produce-to-order a production scheduling system where customer orders precede the production of the item (i.e., the manufacturer must have an order in hand before issuing production orders to produce the item).

Produce-to-stock a production scheduling system where the production schedule is based on forecast sales and/or actual demand, and the finished inventory is placed in stock for subsequent customer usage.

Productivity cycling process a logic structure used to explain how just-in-time techniques can lead to market dominance in any industry.

Repetitive manufacturing a type of production where discrete units of a fairly homogeneous product are produced at relatively high speeds and volumes. Also characterized as *continuous flow production* in a sequentially organized operation.

Seiri a Japanese term used to represent the proper arrangement of work stations, inventory, and work activities.

Seisetsu a Japanese term used to remind workers to maintain their work station's equipment and tools.

Seiso a Japanese term used to represent cleanliness of work stations.

Seiton a Japanese term used to represent orderliness of work stations, inventory, and work activities.

Shitsuke a Japanese term used to remind workers to follow rules and make waste removal a habit.

Total quality control (TQC) a concept or approach to performing a quality assurance program in a just-in-time operation.

Work-in-process (WIP) inventory in various levels or stages of production completeness. This would include inventory from raw materials up to finished products awaiting inspection and acceptance as completed.

References

[1] K. A. Wantuck, *Just-In-Time For America: A Common Sense Production Strategy*, The Forum, LTD., Milwaukee, WI, 1989.

[2] R. J. Schonberger, *World Class Manufacturing: The Lessons of Simplicity Applied*, The Free Press, New York, 1986.

[3] H. Hirano, *JIT Factory Revolution: A Pictorial Guide to Factory Design of the Future*, Productivity Press, Cambridge, MA, 1989.

[4] R. J. Schonberger, *Japanese Manufacturing Techniques: Nine Hidden Lessons in Simplicity*, The Free Press, New York, 1982.

[5] A. S. Sohal, A. Z. Keller, and R. H. Fouad, "A Review of Literature Relating to JIT," *International Journal of Operations and Production Management*, Vol. 9, No. 3, 1989, pp. 15-25.

[6] B. Dutton, "Switching to Quality Excellence," *Manufacturing Systems*, Vol. 8, No. 3, 1990, pp. 51-53.

CASE 1

JIT'ing The Wrong Way

Bacon Processing Plant (BPP) of Sheepdip, Kansas, is an electronic components manufacturer with a single production facility of 125,000 sq. ft., employing 2,000 assembly line workers, 1,200 skilled workers and 850 management and clerical staffers. The upper management of BPP attended a seminar on JIT offered by their major customer, the U.S. government. After some discussion, upper management sent down a dictum to first line supervisors (plant foremen) that BPP was going to become a JIT operation.

BPP manufactures electronic components for the U.S. government. Their components are used as replacement parts in military equipment and office equipment. The government purchases the products in large lots for their own disbursement reasons. The government currently has no plans to change their lot size orders in the near future. The U.S. government's business represents about 80 percent of total sales for BPP.

In recent months the government inspectors noticed that BPP's JIT-oriented competitor's products were improving in quality, while BPP's quality by comparison decreased. BPP's past acceptance sampling methods simply did not catch all of the defective products, while similar products from the competitor were virtually defect free. Also, a growing number of new customers from industry had in recent months placed pressure on BPP to adopt JIT methods that would permit a quicker response from BPP on orders. The new customers expressed a desire to have smaller lot sizes shipped to them instead of the typical large lot size the BPP system had been designed to provide because of past government business. What's more, a foreign competitor recently entered the market and could provide a better quality product, at less cost, and in a more timely manner than BPP.

The task of learning what JIT was, devising an implementation plan, and implementing it was left entirely up to the first line supervisors. BPP upper management felt that the supervisors managing the workers who were going to be asked to perform JIT activities at the shop floor would be the best people to make the new system work. The management of BPP also realized converting to a more responsive and quality-conscious product system had to be accomplished as soon as possible. They established a deadline for full implementation of JIT principles in ninety days. The foremen and supervisors were given thirty days to develop a plan, thirty more to train and install the plan, and thirty to work out any bugs. As an added inducement, upper management informed the workers that 100 percent of

EXHIBIT 1-1 • *A List of Targeted Activities*

Try to reduce the time it takes to perform these activities since they represent a waste of time and add little to the value of our products:

counting component parts

counting boxes

cutting material used in the product

maintaining equipment

switching machine on

over-producing stock

handling materials

Try to do a better job, even if it takes more time, in performing the following activities since they add value to our products:

making sure components fit in subassemblies

checking other workers' jobs

checking orders to see they are correct

cleaning up work centers

moving boxes

setting up machines for a production run

---

increased productivity JIT benefits caused by their cost-reducing would be given to them as a reward.

The supervisors felt that much of the JIT philosophy was based on identifying time-consuming, wasteful activities and reducing them where possible. At the same time, work activities that added value to the product should be identified, and where possible, have a renewed effort or an increased allocation of time invested in them. The first act by the supervisors was the publishing of the list presented in Exhibit 1-1. This list was designed to identify targeted areas of improvement or waste removal.

Case Questions

1. Do you think the BPP assignment to the first line supervisors to implement the new JIT program was a correct move by upper management? Explain your answer.

2. What JIT key principles does the application in this case appear to violate? Which principles does it embrace?

3. Which of the items listed in Exhibit 1-1 should be listed as "motion time" activities that should be eliminated? Which should be listed as "work time" activities that add value to the product?

4. Do you think this JIT program will be successful? Give reasons why or why not.

CHAPTER TWO

JIT Inventory Management

Chapter Outline

Introduction

JIT Inventory Management Principles

Comparison of JIT and Large-Lot EOQ Operations

JIT Purchasing and Supplier Relations

Characteristics of a JIT Partnership

New Technology Supports JIT

A Methodology for Implementing JIT in an EOQ Environment

Methodology Assumptions

JIT/EOQ Models

An Illustrative Example

A Methodology for JIT Vendor Evaluation

Objective Evaluation Criteria Related to Vendor Quality

Evaluation Methodology for Use with Objective and
Subjective Criteria

Measuring JIT Inventory Management Performance

Implementation Strategy for JIT Purchasing

Learning Objectives

After completing this chapter, you should be able to:

1. Explain what just-in-time (JIT) inventory management principles are and how they can improve a manufacturing operation.
2. Explain the differences caused by the differing inventory principles between a typical EOQ operation and a JIT/EOQ operation.
3. Define the characteristics of JIT purchasing.
4. Describe a new ordering/purchase technology used to support the implementation of JIT.
5. Explain and be able to use JIT/EOQ modeling methods to minimize inventory costs in an EOQ environment.
6. Explain and be able to use a scoring method to evaluate JIT vendors.

Introduction

The key principles presented in the last chapter only begin to establish a framework for the process called just-in-time (JIT). To give substance to this framework a closer examination of various operating functions within an operation must be undertaken. In the next chapter, Chapter 3, we will explore the impact of JIT on production planning and scheduling management functions. In Chapter 4 the impact of JIT on the quality control function will be examined. The focus of this chapter will be to examine how JIT impacts the inventory management function.

Inventory management can be generally defined as the management function of planning and controlling inventory. The inventory includes raw materials, component parts, subassemblies, work-in-process (WIP), and finished goods. Inventory management involves the securing of inventory for a production facility (usually from a supplier or vendor), allocation within the facility for its use in the product being produced, and its storage until required by customers. All of the policies, rules, procedures, methodologies, and principles that are related to inventory and its flow through the production facility are a part of inventory management. A JIT operation has its own set of inventory policies, rules, procedures, methodologies, and principles that are a part of the key JIT principles presented in Chapter 1.

Inventory management principles of JIT are outlined in this chapter. In addition, a comparative analysis is presented to illustrate the differences between a JIT operation and a large-lot "economic order quantity" (EOQ) operation. Several inventory measurement methods are discussed as a means of evaluating JIT performance. The key role of the supplier is also discussed, along with some of the more recent modeling formulas used in determining lot size ordering and vendor performance [1, 2, 3].

JIT Inventory Management Principles

There are a great many inventory management policies, rules, and procedures that are a part of JIT [4]. Six of the more commonly used of these can be characterized as JIT inventory management principles. These principles include:

1. Cut lot sizes and increase frequency of orders
2. Cut buffer inventory
3. Cut purchasing costs
4. Improve material handling
5. Seek zero inventory
6. Seek reliable suppliers

1. Cut Lot Sizes and Increase Frequency of Orders In a JIT operation the ideal lot size is one. Remembering the unitary lot size production goal from Chapter 1, we need to have all **dependent demand inventory** (i.e., raw materials, component parts, subassemblies, etc.) arrive just in time for their use. Since we use dependent demand inventory like component parts one-at-a-time, doesn't it make sense that we should plan to use this type of inventory one-at-a-time? Why have a lot of 100 units of a component part arrive at one time, when only one part is necessary at any one time?

When lot sizes are cut, workers at work centers do not have as much idle inventory stored around them. Idle inventory involves a wasteful cost of capital investment, wastes costly space in the facility, and can become a physical obstacle reducing productivity at work stations. Cutting lot sizes, while increasing the frequency of orders to balance the demand needs, reduces costly waste and improves productivity.

Cutting large-lot sizes also involves the **independent demand inventory** or finished goods inventory. In a JIT operation, the orders for finished goods are sent to other JIT customers in small, but frequent, lot sizes. This saves costly space in the shipping department that would be required to stage large-lot size orders, reduces or eliminates the need for warehouses, and increases productivity by reducing inventory clutter and material handling efforts.

2. Cut Buffer Inventory Idle inventory in an inventory department or in WIP becomes a costly **buffer stock** (also called a *safety stock*) that prevents problems from revealing themselves.

Suppose a supplier sends substandard quality component parts to a company whose work centers have a week's worth of inventory piled up around them. In this situation it will take a week for the workers to work their way down to the subquality parts and identify that they have a problem. In a JIT operation with the ideal lot size of one and no buffer stock the defect would have been found in the very next assembly of a part. The faster a problem is found, the faster it can be solved and inventory flow improved.

Suppose a worker doesn't know how to assemble several parts together correctly. With a buffer stock at the work station, workers will have the extra buffer stock to work from even if they have to scrap several parts to learn how to do the job right. In a JIT operation with no buffer stock to turn to, workers will have to bring the scrap work to the attention of management in order to complete their next assembly unit. The workers' inability to complete their job can be identified and corrected, resulting in reduced scrap and rework.

Other problems that impede inventory material flow will also be forced to surface. Problems with the poor engineering of parts, malfunctioning of automated equipment, and material handling and delivery flow problems causing delays will all surface quickly. Without a buffer inventory as a back-up, problems relating to inventory flow are motivated by this JIT inventory principle to surface for a speedy solution.

3. Cut Purchasing Costs Increasing the frequency of orders can increase the fixed order charges suppliers and vendors require customers to pay. Having smaller lot or order sizes reduces the possibility of taking quantity discounts and increasing product costs. In addition, JIT's unitary use of inventory materials usually requires special unitary packaging that can also increase purchasing costs. How can a JIT operation hope to cut purchasing costs? There are many ways to cut purchasing costs for JIT operations, starting with the suppliers. JIT manufacturers seek to cut the number of suppliers down to as few as possible. They seek suppliers whose business the JIT manufacturer will dominate, so pricing and service are strongly controlled by that manufacturer. Long-term contracts are negotiated that will allow some flexibility in ordering (i.e., usually an upper and lower bound on possible daily, weekly, or monthly units ordered). The long-term contractual nature and control by the manufacturer can greatly reduce those purchasing cost factors that might increase when using JIT. At the same time, a JIT operation reduces red tape by reducing the number of suppliers and the resulting paperwork. Smaller order sizes can also help cut the formal paperwork required of large-lot size shipments.

4. Improve Material Handling Inventory items coming to a JIT operation from a supplier or vendor should be broken down into the unit or lot size required by the operation. Any imbalance between the arrival quantity at manufacturing plants and their need at that facility will result in unwanted waste. Any imbalance between units of inventory arriving at work centers and their use at work centers will result in unwanted waste. Any imbalance between the shipments to customers and the demand required by customers will result in unwanted waste. The ideal goal in a JIT system is to have no handling by locating the feeder and user processes of materials next to one another.

In a JIT operation we want inventory to arrive in the exact quantity required (ideally on a unitary basis) and just in time for its use. To accomplish this, plant facilities are streamlined to minimize the amount of material handling. (We will talk about plant layout in Chapter 3.) By structuring plants in a version of a **flow shop,** the layout maximizes the flow of production and materials used. Where possible, automated systems augment the layout and are used to replace humans in routine or boring material handling jobs. Robots are used to save workers from material handling, back straining, lifting, and moving tasks. Optical scanning equipment, **bar code** labeling methods, and robots can perform the wasteful activities of inspection, classification, and storage of inventory. The idea in JIT automation is not to eliminate the human workers, but to save them for intelligent work that can take full advantage of their combined mental and physical capabilities.

The frequent but small order quantities JIT inventory principles necessitate have lead to improvements in material handling methods. One popular method devised to support JIT principles is the **reusable container.** Consisting of hard plastic, wood, or metal, reusable containers are being used to reduce packaging waste and protect inventory materials. Shaped in the form of easy-to-stack open buckets, unitary or multiple inventory items can be easily placed into and quickly obtained from these containers by workers performing material handling functions. The added cost of returning the reusable container is usually more than made up for by the fact that their stronger construction minimizes handling damage. More importantly, these containers aid material handling by saving wasteful repackaging. The containers can be used as temporary storage units that permit the contents to be brought directly into the facility and directly to the work centers without the need for repackaging. They also permit the inventory to be loaded directly from a delivery truck into **automated storage/automated retrieval (AS/AR) systems** without repackaging when items are to be warehoused.

Even basic material handling equipment like the truck can be improved to support a JIT inventory system. Instead of the truck door being located in the back, it can be more efficiently moved to the side. Why? These "side-loader truck" doors permit the driver to pull right up to a loading dock without the time-consuming backing-up process. Also, the unloader isn't just limited to a working space at the back end of the truck which can reduce productivity. Instead, he or she can enter the load from the side and can work toward the front, back, or in both directions. Carrying portable ramps under a truck's carriage can also permit greater flexibility in making deliveries to loading docks that don't match up with the level of the truck's doors.

5. Seek Zero Inventory Inventory anywhere is a costly waste of time, effort, and money. Idle inventory sitting in departments or on the shop floor should be eliminated. Inventory being transported is also wasteful, as previously stated in Chapter 1's JIT principles. This leaves only one alternative: there should be zero inventory in a JIT operation. While this may sound like an impossible principle, it is clearly the goal we should seek if we want continuous improvement in inventory cost reduction. Inventory should be reduced or eliminated whenever possible to reduce unwanted waste in the operation.

6. Seek a Reliable Supplier The key to making JIT work is having the inventory just in time for its use. If a supplier's delivery lead time is not reliable, a JIT system will suffer wasteful delays and downtime. In a JIT operation fewer suppliers will be counted on for a greater job of supplying. While a long-term contract and the larger proportion of business the manufacturer represents to a supplier will be helpful in controlling supplier behavior, it cannot always ensure their deliveries will be on time. Some suppliers might have to move closer to their manufacturing customers on a geographic basis to ensure reliability. Other suppliers might have to engage in dedicating some of their contracted carriage activities to JIT manufacturers, such as dedicating a truck or two for just a specific JIT manufacturer. While these practices may seem costly to the supplier, they may be quite appropriate if the majority of the business a supplier does is with a single JIT manufacturer.

The role of the supplier is one of the most important for the success of a JIT operation. We will examine this role more closely in the later section on JIT Purchasing and Supplier Relations.

While the six inventory management principles above are basic to most JIT operations, they do not all have to be applied for an operation to be considered a JIT operation. Few JIT operations will ever really achieve a zero inventory level or even find a completely reliable supplier. The closer an operation is to fully implementing these six JIT inventory management principles, the better the chances are they will receive the many JIT benefits. The JIT benefits that are commonly reported when comparing a JIT operation with a large-lot EOQ operation are the subject of the next section.

Comparison of JIT and Large-Lot EOQ Operations

The JIT inventory management principles above call for a dramatically different manufacturing operation than the classic U.S. large-lot EOQ operation. (They can dramatically alter service operations too, and we will be discussing this in Chapter 7.) To illustrate these differences and elaborate on the JIT inventory management principles, a comparison between a typical JIT operation and a typical large-lot EOQ operation is worth examining.

In Figure 2-1(a) a typical JIT operation employing the JIT inventory management principles is presented. In Figure 2-1(b) a typical large-lot EOQ operation is presented. Both operations produce a finished product "ABC" from three components, "a," "b," and "c," that are obtained from external suppliers. The components are processed as WIP components "A" and then into "AB" and "ABC" in three **work centers** that comprise the manufacturing area of the production facility. The finished products ABC are shipped to retail stores. The areas allocated to the suppliers, manufacturing facilities, and retail stores are proportioned on the basis of need to accomplish the same level of production. In other words, the JIT operation in the figure is producing at the same level as the large-lot EOQ. The additional size of the EOQ operation is required to support the larger quantity of inventory required in a large-lot operation

FIGURE 2-1 • *Comparison of JIT and Large-Lot EOQ Operations*

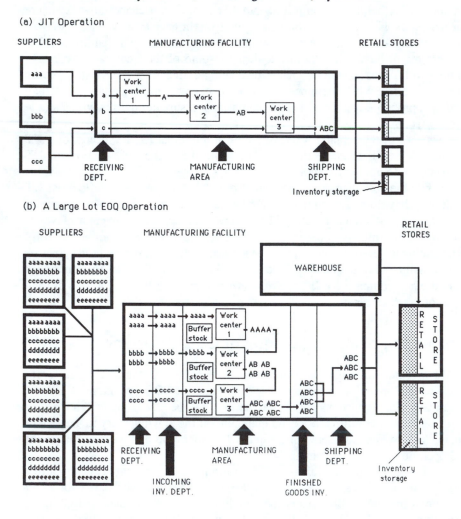

(a) JIT Operation

(b) A Large Lot EOQ Operation

The JIT suppliers are fewer in number and size. The proportion of business the manufacturer does with each supplier permits considerable control of service. The EOQ operation on the other hand has little control when dealing with large suppliers and can afford to occasionally disappoint a small customer. While the JIT manufacturer risks a lack of competitive pressure by dealing with fewer suppliers, it can exert greater pressure because of the volume purchased. The JIT manufacturer also avoids the red tape of dealing with the large quantity of suppliers facing the EOQ operation.

In the JIT operation, small unitary orders are sent to the manufacturing facility frequently. These items are sent directly to the work centers where they are used in production. While the JIT facility in Figure 2-1(a) is structured in a typical flow shop layout, it could be altered (and probably would be) to have the loading doors located nearer to the work centers to reduce material handling efforts. In the EOQ manufacturing facility the large-lot size requires more space in the receiving department to receive and stage the order for inspection. A large lot might be the size of a **pallet** or twenty pallets. When large orders come in, the possibility of shortages is more easily hidden and therefore must be more carefully checked than the single item deliveries of the JIT operation. The EOQ operation also temporarily stores its inventory in the incoming inventory department until required for later production requirements. This incoming inventory supply cushions the manufacturing facility from possible shortages caused by supplier delays. There is no need for an incoming inventory department in a JIT operation, as the materials arrive just in time for their use in production. Also, the JIT principles actually invite shortage problems to reveal themselves so they can be corrected and material flow improved. In the actual EOQ manufacturing area, buffer inventory is used to permit workers to keep working even if the WIP has not arrived when they are ready for it. Again, the JIT operation seeks to reduce or eliminate buffer inventory to help find problems in material flow.

When a unit of product is finished in the JIT operation it is sent directly to the customer that ordered it. Under the **produce-to-order** JIT principle, the demand for the item has to exist before it is produced. The retail stores operating with JIT suppliers tend to be smaller than their EOQ counterparts because they do not have to store inventory since the customers are supposed to be waiting for it. This reduces the costs of storage space for the retail store and permits a greater number of retail outlets per JIT manufacturing facility. In the **produce-to-stock** EOQ operation, the production goes into the finished goods inventory department until it is either shipped to a warehouse for more storage or shipped to a retail store. Since an EOQ operation obtains its inventory from suppliers in large lots, it tends to produce and ship orders to customers in large lots. This results in the need to have warehouses to store excess production (manufacturing facility storage tends to be very costly) or large retail store inventory storage space.

TABLE 2-1 • *A Summary of Differences Between EOQ and JIT Operations*

Suppliers	*EOQ*	*JIT*
Size	large	small
Number	many	few
Geographic location	dispersed	near manufacturing facility
Manufacturer control over changes in:		
price, lead time, quantity	little	a lot
Order quantity size	large lot	small or unitary lot
Receiving Dept.	*EOQ*	*JIT*
Space	a lot	little
People	a lot	little
Equipment	a lot	little
Info. processing	a lot	little
QC inspection	some	little or none
Incoming Inv. Dept.	*EOQ*	*JIT*
Space	a lot	little or none
People	a lot	little or none
Equipment	a lot	little or none
Info. processing	a lot	little or none
Auditing	a lot	little or none
Manufacturing Area	*EOQ*	*JIT*
Space	average	above average
People	average	above average
Buffer stock area	a lot	little or none
Job tasks	few	many
Job training	narrow, focused	broad, flexible, crosstrain
QC tasks req. of workers	little or none	many
Control on line stoppage	little or none	a lot
Production rate	forecast pushed	demand pulled quota
Simple QC checks performed by	workers	robots
Qual. Cont.(QC) Dept.	*EOQ*	*JIT*
People	a lot	little or none
People use	inspection	idea facilitator and education
Level of automation	low	high
Procedure for locating defects	sampling	100% inspection

Finished Inv. Dept.	*EOQ*	*JIT*
Space	a lot	little or none
People	a lot	little or none
Equipment	a lot	little or none
Info. processing	a lot	little or none
Auditing	a lot	little or none
Shipping Dept.	*EOQ*	*JIT*
Space	a lot	little
People	a lot	little
Equipment	a lot	little
Info. processing	a lot	little
Auditing	a lot	little or none
Warehousing	*EOQ*	*JIT*
Space	a lot	little or none
People	a lot	little or none
Equipment	a lot	little or none
Info. processing	a lot	little or none
Auditing	a lot	little or none
Retail Stores	*EOQ*	*JIT*
Size	large	small
Number	few	many
Inventory storage area	a lot	little
Storage based on	forecast	known demand

In summary, large lots require more space and time to process. More space and time requires more people and facilities to complete the same job that a JIT operation can do with less. The difference between the two represents wasted resources that if saved could result in a more productive and successful operation. A detailed summary of these differences (some of which we will discuss in other chapters) is presented in Table 2-1.

JIT Purchasing and Supplier Relations

In the classic large-lot operation, purchasing decisions are based on the cost-minimized EOQ formulas that define how many units of inventory to order and when the orders should be placed. These EOQ formulas are usually expressed as a function of carrying costs (e.g., insurance, storage space, taxes, etc.) and ordering costs (e.g., fixed order charges, expediting, purchasing department personnel, etc.). As presented in Figure 2-2(a), the intersection of the carrying cost and ordering cost functions define the cost-minimizing EOQ (i.e., EOQ_1 in Figure 2-2a). Many U.S. organizations have for decades based their inventory systems on EOQ modeling. Of those who have moved from EOQ to JIT, many have chosen a logical path of slow and gradual movement from a large-lot order size to a smaller JIT lot size. This is not only in keeping with the JIT inventory principle of seeking smaller lot sizes, but the JIT system actually helps to motivate the change to take place [5]. Reductions in whole classes of carrying costs are started by adopting smaller JIT lot sizes and methods. In Figure 2-2(b) we can see that a reduction in ordering costs causes EOQ_1 to be reduced to EOQ_2. As we seek longer-term contracts with reduced fixed ordering costs and reduce the number of suppliers (hence, the red tape in keeping track of them), the EOQ shifts downward as in Figure 2-2(b). The full impact of the JIT inventory principles, though, results in still further shifts in the EOQ. As the lot size is reduced, so is average inventory.

The average inventory of the original EOQ is greater than the average inventory of a reduced EOQ. This will cause a reduction in carrying cost since many of the carrying cost items (like taxes or insurance) are based on average inventory. The result of reduced carrying cost causes a shift increasing the EOQ as presented in Figure 2-2(c). The complete effect of a round of cost reductions will net a reduction in the EOQ from EOQ_1 to only EOQ_3. With the JIT principle of continual improvement and cost reduction at work, implementing the JIT system will help to achieve its own stated goal of an EOQ of one. To help marshal this built-in motivational force and direct it towards a successful JIT implementation requires a successful partnership between JIT purchasing and suppliers.

Characteristics of a JIT Partnership

JIT requires a special relationship between the supplier and the purchasing department of a JIT facility [6, 7]. The relationship between the supplier and purchaser must be one of a cooperative partnership where both parties work together to accomplish a prosperous future. Some of the characteristic features of this relationship include:

1. Long-term contracts.
2. Improved accuracy of order filling.
3. Improved quality.
4. Ordering flexibility.
5. Small lots ordered frequently.
6. Continuous improvement in the partnership.

Collectively, these characteristics represent what is called JIT purchasing. Let's look at each as guides to expected behavior between the supplier and purchaser in a JIT operation.

FIGURE 2-2 • *Impact of JIT Cost Reductions on Lot Size*

(a) Basic Large Lot EOQ Cost Graph

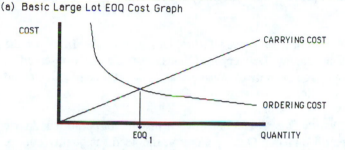

(b) Impact of JIT Reduced Ordering Cost

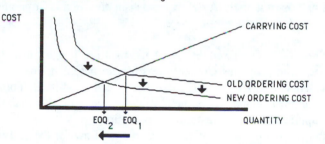

(c) Subsequent Impact of JIT Reduced Carrying Cost

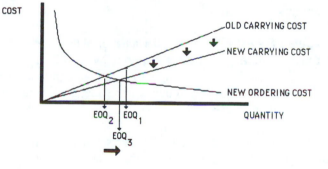

1. Long-term Contracts In a JIT operation, demand determines the purchasing decisions of order quantity and timing. The long-term nature of the contract is designed to provide the suppliers with the security of knowing they will not be dropped and stuck with unwanted inventory. It is not uncommon for the contract to have a single-source stipulation, so the supplier will be the sole supplier of the inventory item to the JIT purchaser. The security of the contract for the supplier should be used to justify a reduced unit cost and ordering cost (e.g., the fixed charge to place an order). The long-term nature of the contract for the purchaser is designed to provide some leverage in controlling prices, quality, and delivery lead times.

2. *Improved Accuracy of Order Filling* Orders must be filled by the supplier without error in quantity. Delivery lead times must be exactly observed. A failure to deliver the required quantity of inventory at the exact time required could bring a JIT operation to a stop.

3. *Improved Quality* Orders with defective items are not permitted. As we will discuss in Chapter 4 on Total Quality Control, the need for quality control inspections of incoming materials should be reduced or eliminated by the elimination of defects in those materials. A defect from a supplier in a JIT operation causes a shortage that in turn will cause a stoppage.

4. *Ordering Flexibility* The fixed EOQs of the past are history in a JIT operation. The JIT operation produces to meet the shifts in actual demand, not forecast demand. As demand shifts, so will there be a need to shift inventory requirements used to meet the demand. The more closely we match production with inventory flow from suppliers, the less waste in the operation we will experience. The contractual requirements must be sufficiently flexible to allow for the daily or even hourly shifts in inventory ordering. Communication systems must also be in place to provide supplier and purchaser a quick and easy process of dialogue in periods of demand volatility.

5. *Small Lots Ordered Frequently* The supplier must be able to provide the smaller and more frequent ordering quantities required by a JIT operation. Moreover, the supplier must be flexible enough to allow the purchaser to continually shift to lower lot sizes as they move toward the goal of unitary production where the order quantity equals one.

6. *Continuous Improvement in the Partnership* The supplier is expected to work with the purchaser in helping to reduce material unit costs to the purchaser, seeking new cost reductions in material handling and shipping to the purchaser, and working to solve jointly new unitary packaging problems, solving shipping problems and improving material quality control problems. The purchaser is not only expected to live up to long-term contracts but work with the supplier on all joint problems. In addition, the purchaser should provide information to the supplier on the state of JIT implementation and how the supplier's efforts are helping or hurting the purchaser's success. The purchaser must also work to establish and use communication systems to keep the partnership active and informed. Joint efforts to invest in equipment that may facilitate loading and

unloading of supplier and purchaser trucks is just one area where tangible cooperation and communication between the partners can increase the productivity of both.

The benefits of these characteristics include lower inventory carrying cost, reduced scrap and rework, improved finished goods quality, reduced quality control, reduced inspection, faster response to order changes, and a reduction in all purchasing department resources. In other words, the use of JIT purchasing with a successful supplier can substantially reduce the waste of resources and improve productivity.

New Technology Supports JIT

In the late 1980s a new technology appeared in industry that is ideal for supporting the communication and flexibility characteristics of a JIT operation. This new technology is called **electronic data interchange (EDI)**. EDI is used to exchange supplier and purchaser orders directly between companies' computer systems [8]. The EDI system enhances the purchasing and supplier relationship by minimizing the amount of time it takes to receive, process, and make order changes.

EDI requires three elements: a message standard, translation software, and a communication network. A **message standard** is a predefined standardized format for all types of purchase orders that companies plan to send each other. While the purchasing company sending an order will have its own order form, the supplier taking the order will have its own invoice form on which to record the order. Both forms will contain basically the same information with the exception of some unique types of information that are only used for internal processing of the order (e.g., a bin number for the supplier to locate the purchased items).

The **translation software** element of EDI is used to interpret the purchasing company's order message into the supplier's order form, despite the fact that both forms may contain unique information and different formats. The translation software can actually reach into an organization's computer based ordering system and retrieve or change order information on a more timely basis than waiting for the purchasing agent to implement such changes manually.

Finally, the **communication network** is used to transmit the purchase orders. The most common communication networks that support the EDI system are electronic mail boxes. An **electronic mail box** is a computer terminal or computer at a work center that can receive and send messages to other terminals or computers over telephone lines.

The EDI system works to speed the sending, receiving, and processing of orders between supplier and purchaser. Purchase orders are electronically entered at a terminal at the purchaser's facility and sent to another terminal at the supplier's facility via the telephone lines. The supplier's computer stores the order until the supplier accesses it. The supplier may need only to press a single return key to authorize an order to be picked. In more fully computer integrated systems the purchaser may be able to receive an immediate confirmation on the status of the order when it is entered on the terminal. In turn, the supplier would be able to send the status of the order, including any stockouts, right back to the purchaser via their electronic mail box in a matter of minutes rather than days.

The EDI system has been observed to support JIT by reducing the time that it takes for the purchasing company to send an order, since the order does not have to be carried through the mails, but instead is delivered at the speed of electricity to the supplier by long distance phone lines. In addition, the invoice preparation time by the supplier is reduced or eliminated since the translation software converts the purchase order from the purchasing company into an invoice order for the supplier. Other benefits of this technology that support JIT principles include improving order accuracy and order taking productivity (the order is not rewritten by humans), facilitating purchaser-supplier communication, and reducing wasted paper flow, document storage, and filing costs (EDI messages can be entirely electronic).

A Methodology for Implementing JIT in an EOQ Environment

Many manufacturing operations in the U.S. still receive and send inventory in large-lot sizes. Although JIT has been encouraging lot sizes toward ever smaller amounts, little in the way of formal methodology has been offered as a guide. Most of the reductions in lot size have been made experimentally.

Since many of the firms moving toward a JIT system are doing so from an EOQ modeling environment, it might be beneficial to use the EOQ modeling approach to help make the transition to JIT. Most all inventory managers understand and still appreciate the logic of EOQ modeling. EOQ modeling can be used to demonstrate JIT's cost-reducing benefits to managers making the change to a JIT operation [9]. Moreover, new JIT-based models can be used to suggest order quantities and the number of deliveries a manufacturer should receive under the long-term JIT purchasing contracts.

In the basic cost minimizing EOQ model (i.e., just minimizing carrying and ordering costs), models are used to determine order quantity and total annual cost. These basic EOQ models are presented here.

Equation 2-1.

$$1.\ \text{EOQ ORDER QUANTITY}\ (Q^*) = \sqrt{\frac{2\,O\,D}{C}}$$

Equation 2-2.

2. EQO TOTAL ANNUAL COST(T^*) = CARRYING COST + ORDERING COST

$$= \frac{C\,Q^*}{2} + \frac{O\,D}{Q^*}$$

Where: Q^* = cost minimizing order quantity in units under EOQ system
O = ordering costs in dollars per order
D = annual demand in units
C = carrying costs in dollars per unit
T^* = minimized total annual cost in dollars under EOQ system

The optimized EOQ value of Q^* of Equation 2-1 represents the fixed order quantity lot size that an organization would place each time an order is necessary to replenish stock. An **order point** model is also required for the EOQ modeling system to indicate when the next order of size Q^* should be placed. There are a great many order point models to choose from, but since the implementation of the following proposed JIT/EOQ system renders them unnecessary, we will not be discussing them here. (For a good review on the subject of order point modeling, see Gaither [7, pp. 467-482].)

Methodology Assumptions

The basic EOQ model is often criticized for unrealistic assumptions on which it is based. Let's examine those EOQ assumptions in light of a switch to a JIT operation. Stevenson [1] reports the six assumptions for the basic EOQ model to be:

1. Only One Product Is to Be Considered in the Model In a JIT operation this assumption is even more restrictive. JIT seeks unitary production goals where each unit of a product is considered a unique and separate order.

2. Annual Total Demand Requirements Are Known In a JIT operation nothing is produced until an order has been placed. Yearly, monthly, weekly, daily, or even hourly demand has to be known with relative certainty in a JIT operation.

3. Annual Demand Usage Is Spread Evenly to Achieve a Fairly Constant Usage or Demand Rate from Customers In a high-volume or repetitive JIT operation, demand is expected to have some fluctuation, but in general to be fairly constant. In a low-volume JIT job shop operation, a high degree of fluctuation is often possible, but a large EOQ ordering of inventory would not be appropriate anyway. So a JIT operation also assumes a fairly constant usage rate.

4. Order Delivery Lead Time Is Constant In JIT purchasing we expect the order lead times to be dependable or constant once they are mutually set by the supplier and purchaser.

5. Each Order is Received in a Single Delivery JIT operations seek, where feasible, unitary delivery to support the unitary production. Each order down to the individual item is viewed as a single delivery.

6. There Are No Quantity Discounts In general, the long-term nature of JIT purchasing contracts does not include quantity discounts. Since suppliers usually have to absorb some of the frequent ordering costs required in serving a JIT operation, they tend not to give quantity discounts. Indeed, the principles of JIT motivate purchasers to cut lot size, not increase it for a quantity discount.

In summary, the assumptions used in the basic EOQ model turn out to be basic requirements in a JIT operation. The difference between the inventory approaches of EOQ and JIT is in the use of assumptions when implemented. We can still use (or misuse) the EOQ model when implemented even if the assumptions are unrealistic and not true. When we implement a JIT system, though, the assumptions have to be true or a JIT system cannot be implemented. In other words, a JIT system treats the assumptions as given principles (or given requirements) that must be in place and working before we can call the system JIT.

Based on EOQ formulas, a series of combined JIT/EOQ formulas have been suggested to help bridge the transition from EOQ to JIT [8]. These JIT/EOQ formulas are based on the JIT logic of cutting delivery lot size as a means of implementing JIT in large-lot EOQ environments. For the purposes of this illustration, the following assumptions for the combined JIT/EOQ models must hold:

1. Unit costs are not affected by order quantity.
2. Carrying costs are not affected by order quantity.
3. Ordering costs remain constant no matter how many deliveries are scheduled.

These assumptions are similar to those in the basic EOQ model and are reasonable in light of the control provided to the JIT purchaser in the negotiation of the long-term JIT contracts.

JIT/EOQ Models

The derivation for the JIT/EOQ models in Equations 2-3 through 2-9 can be derived from the total annual cost T^* Formula 2 in Equation 2-2. Using the same calculus and algebraic procedures the originator F. W. Harris used to derive the EOQ formulas, the combined JIT/EOQ models are derived [9]. The difference between the two modeling approaches can be seen by comparing the slight modification of the T^* (Formula 2) and T_{jit} (Formula 4). In changing an EOQ environment to a JIT system, lot size purchasing needs to be cut. This is accomplished by still using the EOQ for optimizing purchase ordering while cutting the Q^* quantities into smaller q delivery quantities. We call this cutting of a large-lot size into smaller quantities for delivery a **split delivery**. The problem in the JIT/EOQ system is to determine the optimal number of n deliveries for the JIT operation based on the original EOQ order quantity. The only difference in the total annual cost functions in the EOQ model and the JIT/EOQ model is the addition of n in the denominator of the carrying cost component of the JIT/EOQ model. From this simple difference, all of the JIT/EOQ models in this chapter are derived.

Equation 2-3 shows the optimal JIT/EOQ order quantity formula. The minimum total annual cost under the JIT/EOQ system can be computed using Equation 2-4 and the resulting split delivery level of q is found using the ratio in Equation 2-5. Since a reduction in average inventory is the logical result in cutting delivery lot size, there is a savings in switching from an EOQ to the JIT/EOQ system. The monetary savings S from the EOQ T^* can be determined using Equation 2-6.

Equation 2-3.

$$\text{3. JIT/EOQ ORDER QUANTITY } (Q_n) = \sqrt{n}\ Q^*$$

Equation 2-4.

$$\text{4. JIT/EOQ TOTAL ANNUAL COST} (T_{jit}) = \frac{C\,Q^*}{2n} + \frac{O\,D}{Q^*} = \frac{1}{\sqrt{n}}\,(T^*)$$

Equation 2-5.

$$\text{5. JIT/EOQ DELIVERY QUANTITY } (q) = \frac{Q_n}{n}$$

Equation 2-6.

6. SAVINGS BY SWITCHING TO JIT/EOQ (S) = $\left(1 - \dfrac{1}{\sqrt{n}}\right)(T^*)$

Where: Q_n = cost minimizing JIT order quantity in units per each "n" delivery

n = optimal number of deliveries during the year

Q^* = cost minimizing order quantity in units under EOQ system

T^* = minimized total annual cost in dollars under EOQ system

q = optimal number of units per delivery

T_{jit} = minimized total annual cost in dollars under JIT system

One of the required inputs into all of the JIT/EOQ formulas above is knowing what the *n* (number of optimal deliveries) should be. Three different formulas for use in determining *n* are presented in Equations 2-7, 2-8, 2-9. In a situation where an *m* maximum unit inventory capacity restriction exists (e.g., physical space limitation, legal regulation on unit storage, tax reasons, etc.), Equation 2-7 can be used to determine the number of deliveries. If a situation exists where a specific on-hand average inventory target is desired, Equation 2-8 can be used to determine the number of deliveries. In a situation where a prespecified percentage of savings in total costs is desired, Equation 2-9 can be used to determine the number of deliveries. Once the number of *n* deliveries is set, the JIT/EOQ total cost T_{jit} and savings *S* from switching to the JIT/EOQ system can be computed.

Equation 2-7.

7. JIT/EOQ Optimal Number of Deliveries (n_m) = $\left(\dfrac{Q^*}{m}\right)^2$

Where: n_m = optimal number of deliveries with "m" maximum inventory capacity limitation

Q^* = cost minimizing order quantity in units under EOQ system

m = maximum inventory capacity level

Equation 2-8.

$$8. \text{ JIT/EOQ Optimal Number of Deliveries } (n_a) = \left(\frac{Q^*}{2a}\right)^2$$

Where: n_a = optimal number of deliveries with an "a" targeted level of
average on-hand inventory

Q^* = cost minimizing order quantity in units under EOQ system

a = a specific targeted average inventory level in units

Equation 2-9.

$$9. \text{ JIT/EOQ Optimal Number of Deliveries } (n_p) = \frac{1}{(1-p)^2}$$

Where: n_p = optimal number of deliveries with a prespecified percentage
of "p" savings in total costs

p = a desired prespecified percentage of total costs savings

An Illustrative Example

To illustrate the JIT/EOQ modeling of order quantities, let's look at an example. An organization is currently using a large-lot EOQ system of ordering units of a single item inventory component. Their product is sold on a contractual basis and yearly demand is known to be 40,000 units. The carrying cost per unit is $1 and the per order cost of purchasing the units is $50. The firm decides that they want to shift their operation to a JIT system. To start the JIT implementation, a long-term JIT purchasing contract has just been signed committing the organization to an overall quantity to be delivered during a period of a year. The overall quantity is based on known demand of the organization's customers and the known production capacities of the manufacturing operation. The supplier is flexible as to when and how many units will be shipped, but requires that once the number is fixed, the order quantity will remain fixed for a designated period of time. The supplier is also willing to make as many deliveries as requested by the purchaser, but again, once the delivery quantity is set, it should remain set for a mutually agreeable period of time. Given this background information, let's look at the following examples:

Example 1.
What are the optimal EOQ order quantity and total annual cost?

As we can see, the resulting EOQ is 2,000 units, which in turn results in a total annual cost of $2,000. In this system the lot size of the delivery quantity is the EOQ of 2,000 units.

$$\text{EOQ ORDER QUANTITY } (Q^*) = \sqrt{\frac{2\,OD}{C}}$$

$$= \sqrt{\frac{2\,(50)(40,000)}{1}}$$

$$= 2,000 \text{ units}$$

$$\text{EQO TOTAL ANNUAL COST} (T^*) = \frac{CQ^*}{2} + \frac{OD}{Q^*}$$

$$= \frac{(1)(2,000)}{2} + \frac{(50)(40,000)}{2,000}$$

$$= \$2,000$$

Example 2.

What would the optimal JIT/EOQ order quantity be if the firm has chosen to reduce the lot size by splitting the order quantity into two deliveries (i.e., $n=2$)?

$$\text{JIT/EOQ ORDER QUANTITY } (Q_n) = \sqrt{n}\ Q^*$$

$$= \sqrt{2}\ (2,000 \text{ units})$$

$$= 2,828 \text{ units}$$

$$\text{JIT/EOQ TOTAL ANNUAL COST} (T_{jit}) = \frac{1}{\sqrt{n}}\,(T^*)$$

$$= \frac{1}{\sqrt{2}}\,(\$2,000)$$

$$= 0.707\,(\$2,000)$$

$$= \$1,414$$

$$\text{JIT/EOQ DELIVERY QUANTITY } (q) = \frac{Q_n}{n}$$

$$= \frac{2,828}{2}$$

$$= 1,414 \text{ units}$$

$$\text{SAVINGS BY SWITCHING TO JIT/EOQ } (S) = \left(1 - \frac{1}{\sqrt{n}}\right)(T^*)$$

$$= \left(1 - \frac{1}{\sqrt{2}}\right)(\$2,000)$$

$$= (0.293)\,(\$2,000)$$

$$= \$586$$

The resulting JIT/EOQ optimal order quantity of T_{jit} increases to 2,828 units and a split q delivery quantity of 1,414 units. The result of lowering the lot size delivery quantity is a total annual cost of only $1,414, down from the EOQ cost of $2,000. This reinforces the JIT principle of how a reduction in lot size can reduce inventory costs. By switching to the JIT/EOQ split delivery of a smaller lot size the organization reduced its cost by $586.

Example 3.

What would the optimal JIT/EOQ order quantity be if the firm has a maximum inventory storage limit of only 500 units?

$$\text{JIT/EOQ Optimal Number of Deliveries } (n_m) = \left(\frac{Q^*}{m}\right)^2$$

$$= \left(\frac{2,000}{500}\right)^2$$

$$= 16 \text{ deliveries}$$

$$\text{JIT/EOQ ORDER QUANTITY } (Q_n) = \sqrt{16} \ (2,000 \text{ units})$$

$$= 8,000 \text{ units}$$

$$\text{JIT/EOQ TOTAL ANNUAL COST} (T_{jit}) = \frac{1}{\sqrt{16}} (\$2,000)$$

$$= \$500$$

The number of deliveries has to be increased to 16 to permit a smaller but more frequent delivery of inventory. The resulting JIT/EOQ optimal order quantity then becomes 8,000 units, resulting in a total annual cost of only $500. This again reinforces the JIT principle of how a reduction in lot size can reduce inventory costs. While the lot size actually increased, the negotiated smaller and more frequent deliveries inherent in JIT purchasing can substantially reduce total inventory costs.

Example 4.

What would the optimal JIT/EOQ order quantity be if the firm has set an average inventory target of having no more than 200 units on hand? The number of deliveries has to be increased to 25 to achieve the 200 unit goal. The resulting JIT/EOQ optimal order quantity then becomes 10,000 units, resulting in a total annual cost of only $400. This again reinforces the JIT principle of how a reduction in on-hand inventory can reduce total costs.

$$\text{JIT/EOQ Optimal Number of Deliveries } (n_a) = \left(\frac{Q^*}{2a}\right)^2$$

$$= \left(\frac{2,000}{(2)\,(200)}\right)^2$$

$$= 25 \text{ Deliveries}$$

$$\text{JIT/EOQ ORDER QUANTITY } (Q_n) = \sqrt{25}\ (2,000 \text{ units})$$

$$= 10,000 \text{ units}$$

$$\text{JIT/EOQ TOTAL ANNUAL COST}(T_{jit}) = \frac{1}{\sqrt{25}}(\$2,000)$$

$$= \$400$$

Example 5.

What would the optimal JIT/EOQ order quantity be if the firm has prespecified a desire to achieve a 30 percent total cost savings from the original EOQ T^* of $2,000?

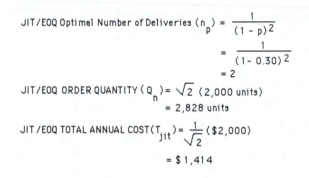

$$\text{JIT/EOQ Optimal Number of Deliveries } (n_p) = \frac{1}{(1-p)^2}$$

$$= \frac{1}{(1-0.30)^2}$$

$$= 2$$

$$\text{JIT/EOQ ORDER QUANTITY } (Q_n) = \sqrt{2}\ (2,000 \text{ units})$$

$$= 2,828 \text{ units}$$

$$\text{JIT/EOQ TOTAL ANNUAL COST}(T_{jit}) = \frac{1}{\sqrt{2}}(\$2,000)$$

$$= \$1,414$$

The computations show that the number of deliveries necessary to achieve the 30 percent cost reduction goal is 2.0408, or rounding to an n_p of 2. The resulting JIT/EOQ optimal order quantity then becomes 2,828 units, resulting in a total annual cost of $1,414. Because the number n_p of 2 deliveries was rounded down to 2, the cost reduction is slightly less than the desired 30 percent, or only 29.3 percent (i.e., $1,414/$2,000). This again reinforces the JIT principle of how a reduction in inventory delivery or lot size can reduce total costs.

The formulas in this section may be helpful in showing how EOQ ordering can be improved by moving to the smaller and more frequent ordering of JIT purchasing. These formulas should also be viewed only as tactics for moving an operation from the large-lot size EOQ mentality to the JIT principles of zero inventory and unitary production. A JIT operation's ordering policies should be based on actual demand and not a formula.

A Methodology for JIT Vendor Evaluation

The role of the JIT suppliers or vendors is critical if JIT purchasing characteristics are to be successfully adopted. In plain terms, a JIT manufacturer must have a JIT vendor as a partner. Unfortunately, even with the best of partners, there are bound to be occasional problems, which may motivate the need for a change in vendor relations or even the dropping of vendors.

Objective Evaluation Criteria Related to Vendor Quality

In a recent survey of over 100 JIT operations, 46 percent reported that some type of formal supplier or vendor evaluation system was in place to evaluate JIT vendor performance [11]. One researcher suggests that purchaser costs to the manufacturer of poor quality be used as objective vendor evaluation criteria. The purchaser's costs that can be directly related to poor JIT vendor behavior include:

1. Scrap and rework costs including such specific measures as rework dollars due to poor quality, sorting costs, excess inventory costs, processing and handling costs.

2. Receiving inspection costs while JIT usually eliminates most of the incoming inventory inspection, vendors whose past poor quality requires extra inspection incur additional costs.

3. Line stoppage and reject costs including such specific measures as line stoppage time due to poor quality multiplied by a production rate per hour, the number of line rejections, machine and labor downtime, and productivity loss.

4. Customer service costs including such measures as customer returns due to poor quality, missed shipments to customers, average outgoing quality levels, dollar value of customer returns, and the number of service calls during warranty periods.

All of these objective criteria are generally recorded and maintained as a normal part of the marketing or operations management functions. Bringing this information together in the form of a JIT vendor management information system (MIS) should be the objective of modern JIT operations. Computer-based systems designed for tracking vendor performance in a JIT environment are starting to be reported in the literature [12]. One of the benefits reported in using a computer-based JIT vendor performance tracking system is that it facilitates communication on supplier problems and makes negotiating changes in contracts with vendors easier.

Evaluation Methodology for Use with Objective and Subjective Criteria

As an organization moves from an EOQ to a JIT operation, it will probably have to drop some of its current vendors or suppliers. Other organizations will have to evaluate which of a number of new JIT vendors to offer a contract. To perform this analysis, some type of comparative methodology is necessary.

One methodology that has been proposed to evaluate JIT vendors is based on a **scoring method** [13]. In this evaluation method a series of JIT vendor criteria are selected as a basis for comparison. The criteria can be either objective (e.g., dollars, percentages, etc.) or subjective (e.g., an ethics rating, a judgment on vendor commitment, etc.). The vendors are then scored for each of these criteria using some type of ranking or numerical system. The scores are then added together and a ranking of the individual vendors can be assessed. In Figure 2-3 a scoring table is presented that commonly is used in the scoring method analysis. The new vendors with the best rankings are selected, or the old vendors with the least desirable rankings are dropped.

One study on JIT vendors has suggested that selection or decision criteria could be divided into three categories [13]: financial, service, and technical. The financial category criteria include the vendor's financial stability in having a long-term existence, its prices to the purchaser, terms of sale, capital investment, size, and order taking systems. The service category criteria include the vendor's on-time delivery ability, delivery frequency ability, ability to provide exact quantities, willingness to be flexible, responsiveness, cooperative nature, frequency of contact, integrity and ethics, willingness to support a long-term contract, location, communication links, transportation capabilities, and commitment. The technical category criteria include the vendor's overall product quality, research and development efforts, and technical resource capabilities. Specific measures for some of these criteria can be objectively determined. They can then be scored in the evaluation process by transforming the objective dollars, percentages, etc., into a numerical score. Other criteria require a subjective assessment for scoring.

Let's illustrate the steps in one scoring process with an example. Suppose an organization has been going through a change from an EOQ system to a JIT system. It currently is doing business with ten JIT vendors and wants to reduce purchasing costs by dropping two of the vendors. Which two vendors should be dropped? To answer this question we can use one version of a scoring method to evaluate the JIT vendors and decide which are the two least desirable vendors to continue to do business with. The following steps can be used in applying a scoring method:

FIGURE 2-3 • *Scoring Method for Vendor Evaluation*

(a) Scoring Table To Determine JIT Vendor Ranking

JIT VENDOR SELECTION CRITERIA	VENDOR NUMBER				
	1	2	3	. . .	n
Criteria 1				. . .	
Criteria 2				. . .	
Criteria 3					
	.	.	.	.	.
Criteria m					
Summation of Scores					
Resulting Rankings					

(b) Scoring Table For Example JIT Vendor Ranking

JIT VENDOR SELECTION CRITERIA	VENDOR NUMBER									
	1	2	3	4	5	6	7	8	9	10
FINANCIAL: Financial stability	5	7	3	8	6	10	1	3	6	4
Price	4	6	2	8	6	10	2	2	4	2
Size	9	3	5	8	6	8	4	3	2	3
SERVICE: On-time delivery	10	7	8	5	2	3	3	1	1	1
Exact quantities	7	5	6	3	2	2	2	1	3	1
Flexibility	4	5	1	2	1	3	1	3	4	1
Responsiveness	8	5	6	2	6	5	1	2	1	1
TECHNICAL: Quality	2	3	7	1	5	1	5	2	1	2
Housekeeping	3	5	7	1	8	2	5	1	1	1
SUMMATION OF SCORES	52	46	45	38	40	44	24	18	23	16
RESULTING RANKINGS	10	9	8	5	6	7	4	2	3	1

1. Determine the JIT Evaluation Criteria That Are Important for the Organization Each organization will require a unique set of evaluation criteria on which to base a selection decision. All the selection criteria stated above, or many other factors, may be required to evaluate a vendor. For the purpose of this example, let's say the nine JIT vendor selection criteria presented in Figure 2-3(b) are the most important for a successful JIT system in the organization.

2. Determine a Score for Each Vendor's Selection Criteria There are many scoring systems available in the literature. For the purpose of this example, let's use a 1 to 10 score, where 1 is best and 10 is least capable. If a vendor has been doing a great job in the category, we would give a high score of 1 or 2. If a vendor has been doing a poor job we would give a low score of 9 or 10. Some evaluators use a simple ranking to differentiate between vendors, while others use complex conversion formulas to change objective information proportionally (i.e., the cardinal differences in product prices cannot always be proportionally reflected in the ordinal shifts of a ranking system). As we can see in Figure 2-3(b), Vendor 1 received an average score of 5, Vendor 6 received the worst possible score of 10 and Vendor 7 received the best possible score of 1. The same score can be given to different vendors for the same selection criteria (e.g., Vendors 3 and 8 both received a score of 3).

3. Sum the Scores for Each Vendor As we can see in Figure 2-3(b), the sum of the selection criteria scores for Vendor 1 is 52.

4. Determine the Ranking of Each Vendor The ranking of each vendor is based on the summation of its scores. Since the smallest summation of scores (i.e., 1 is best score) would represent the collective preferred score over all the other scores, it should be given a rank of one. As we can see in Figure 2-3(b), the smallest summation of scores is 16, and it is given a rank of one. This means that Vendor 10 is the most preferred vendor out of the ten. We continue to rank the other vendors until each has a rank. If the scores are tied, the ranks should be tied for those vendors.

5. Select the Vendors to Keep or Drop The vendors who have the most preferred ranks should be the ones the organization keeps. Those vendors with the least preferred ranks should be dropped. In this example, we would drop Vendors 1 and 2, whose ranks of 10 and 9, respectively, are the least preferred.

The importance of a good partnership between a JIT vendor and a JIT purchaser is critical and should, therefore, not just be left up to trust. Careful and continual evaluation of the few vendors a JIT operation depends on is a necessary part of a successful purchasing control system in modern JIT operations.

Measuring the JIT Inventory Management Performance

Measurement of JIT inventory management efforts can help to demonstrate how they are improving an operation. These measurements can help management show their progress in using JIT inventory principles and act to motivate continual improvement.

One of the more common measures of inventory utilization is **inventory turnover.** Inventory turnover is the ratio of the dollar costs required in producing sold inventory (i.e., material + labor + overhead) to average inventory in dollars. The dollar costs required in producing sold inventory is called **manufacturing cost of sales** or **MCOS.** The average inventory in dollars is the cost per unit times the average inventory level. This ratio can be computed based on data from a month, a quarter, or a year. The inventory turnover ratio provides an estimate of how many times inventory will rotate or turn over in a year. The greater the ratio, the more the inventory is turning over relative to the on-hand average inventory level in dollars. In other words, the greater the inventory turnover, the more it's being utilized.

There are different inventory formulas depending on the cost data used in the numerator. The dollar costs in the numerator are for sales. Only units that are sold are included in the ratio's numerator costs. This limitation is consistent with the JIT principles of production-to-order and product quality perfection. Only finished and successfully sold items will be included in the numerator. While the average inventory in dollars value of the denominator can be based on any period of time, manufacturing cost information is usually collected for a specific period of time. Depending on when the cost data is collected, one of the following inventory turnover ratios can be used:

$$\text{Inventory turnover} = \frac{\text{One month's MCOS} \times 12}{\text{Average inventory} \times \text{Average cost per unit}}$$

$$\text{Inventory turnover} = \frac{\text{One quarter's MCOS} \times 4}{\text{Average inventory} \times \text{Average cost per unit}}$$

$$\text{Inventory turnover} = \frac{\text{One years's MCOS}}{\text{Average inventory} \times \text{Average cost per unit}}$$

These ratios can be used to measure an entire company's inventory usage, a division's usage, a department's usage, or to evaluate any profit or cost center where cost information is collected.

To illustrate the use of the inventory ratio, let's examine an example problem. Suppose a manufacturing company operating under EOQ has an average inventory level in the first quarter of a year of 3,400 units with an average cost per unit of $150. The accounting department reports that the cost of sales for the first quarter of the same year amounted to $300,000 for labor, $125,000 for material, and $50,000 for overhead. During the following year the company started moving toward a JIT operation. The first quarter's average inventory under the JIT system is only 1,400 units. The manufacturing cost of sales or MCOS for this quarter is $425,000. Has the change in systems improved inventory utilization?

Based on the MCOS-based inventory turnover ratio, the company's inventory utilization has improved, as we can see in the ratios below:

Inventory turnover
(under EOQ system)

$$= \frac{\text{One quarter's MCOS} \times 4}{\text{Average inventory} \times \text{Average cost per unit}}$$

$$= \frac{(\$300,000 + \$125,000 + \$50,000) \times 4}{3,400 \times \$150}$$

$$= \frac{\$1,900,000}{510,000}$$

$$= 3.725 \text{ turns}$$

Inventory turnover
(under JIT system)

$$= \frac{\text{One quarter's MCOS} \times 4}{\text{Average inventory} \times \text{Average cost per unit}}$$

$$= \frac{(\$425,000) \times 4}{1,400 \times \$150}$$

$$= \frac{\$1,700,000}{1,400 \times \$150}$$

$$= 8.095 \text{ turns}$$

The inventory turnover in the JIT system is more than twice as much as under the EOQ system. Thus, the larger the ratio on a comparative basis from period to period, the greater the utilization of the average inventory.

Implementation Strategy for JIT Purchasing

One of the ways to start a JIT inventory management program is to begin implementing JIT inventory management principles in the area of purchasing. To implement JIT purchasing a partnership between the supplier and purchaser must be established through a multi-phased process that will embrace both JIT purchasing characteristics and JIT inventory management principles. One JIT expert suggests the following four-phase implementation strategy [14]:

   1. Commitment   Top management of the organization seeking to implement JIT purchasing must make a conscious goal to establish JIT purchasing as a top priority. Top management can achieve this goal by: (1) making it clear that the goal is for the entire organization (and not just the purchasing department), (2) allocating appropriate human and financial resources (which helps to show top management commitment), and (3) playing a visible role in implementing this phase and the remaining phases of implementation. If top management can gain a commitment from employees, the change toward a JIT system is made easier.

   2. Changing the System   Specific changes in the role of purchasing are necessary to move toward a JIT purchasing operation. Form teams of purchasing employees to identify purchasing activities that do not enhance the **value added** to the product being produced. Drop commonly performed non-value-added purchasing activities such as ordering (vendors should help keep track of what is needed), policing (individual managers should keep track of their own vendor quantity and quality failures), and verifying transactions for paying invoices (accounts payable personnel can perform these tasks more efficiently). Most of these activities are duplications of effort as service for other departments and are more efficiently handled by those other departments. A JIT purchasing department must become more focused on fewer purchasing activities. A JIT purchasing manager should focus on sourcing (finding suppliers and vendors that can support the JIT-type operation), pricing (being knowledgeable to negotiate a fair price) and continuously improving relationships with suppliers.

3. Selection of Suppliers   Once the new role of JIT purchasers is established, suppliers who can support a JIT operation must be selected. The teams to select suppliers should be led by purchasing personnel but also include personnel from engineering, quality control, manufacturing, finance, and other areas of the company. The objective of this phase is to select the best supplier for each commodity that the organization purchases. The best supplier may be a single supplier for all the commodities an organization needs or individual suppliers for each commodity. The best supplier should be determined by an analysis of three criteria: (1) ability to meet desired or required service levels (i.e., units, quality, etc.), (2) supplier's capacity for continuous improvement (i.e., supplier's service must continuously improve to permit the purchaser's service level to improve), and (3) commodity price. These criteria can be evaluated using the previously stated JIT vendor evaluation methods.

4. Building Relationships   Once a supplier or group of suppliers is selected, JIT purchasing requires a continuous effort for improving the relationship between the purchaser and the supplier. A good relationship between the supplier and purchaser revolves around a continuous review of supplier performance of five aspects: (1) the quality of the products supplied and the commitment of the supplier to seek continual improvement in quality, (2) the information flow that enhances speed of response from the supplier to the purchaser, (3) the material flow from the supplier to match the timely needs of the purchaser, (4) the reduction in lead time response to changes in customer demands of purchaser, and (5) a price that reflects the level of fair quality.

These four phases of JIT purchasing collectively represent a strategy for starting up JIT in an organization. These four phases are necessary, both as a starting point for implementing a JIT purchasing program, and to help the organization-wide implementation of JIT programs in inventory, production planning and scheduling, and quality control. By structuring inventory flow systems from external suppliers to match an internal JIT production system, the purchaser will receive maximum JIT benefits in both inventory and production activities.

Summary

In this chapter we examined JIT inventory management principles and methodologies. We have seen that inventory managers in a JIT operation should seek to cut lot sizes and buffer inventories in an effort to achieve a zero inventory level. Inventory managers should also try to find reliable suppliers who will work with them to improve material handling and reduce purchase costs. For organizations that are currently switching from an EOQ ordering system to a JIT system, we presented several new JIT/EOQ formulas to guide them in order and delivery quantity decisions. Finally, we also discussed how the improvements caused by the implementation of these JIT principles should be measured using inventory turnover measures, and we presented a methodology to help evaluate vendors in the performance of their JIT services to a manufacturer.

Inventory management is only one function of a production/operations manager. Once the inventory is brought to a JIT operation, it must be converted into the desired products sought by the customer. To accomplish this, production management functions of planning and scheduling must be performed. The subject of the next chapter is JIT production management.

Important Terminology

Automated storage/automated retrieval (AS/AR) system an inventory storage system usually controlled by a computer system incorporating automatic loading and unloading capabilities.

Bar code a label placed on inventory items that is made up of a series of alternating bars and spaces that is encoded by computers.

Buffer stock a planned quantity of inventory stock allocated to work centers or inventory stocking areas as protection against shortages. Also called a "safety stock," this inventory is used to cushion an operation against surges in demand on inventory items.

Communication network an EDI type of communication system that can be used to support computer-based messages.

Dependent demand inventory inventory items that are used to complete, but do not include, a finished product. These items would include such inventory items as raw materials, component parts, and subassemblies. They are usually considered dependent on "independent demand" inventory or finished goods.

Economic order quantity (EOQ) an order quantity that minimizes total carrying and ordering costs. The EOQ is derived by modeling cost functions of individual inventory items.

Electronic data interchange (EDI) computer software that permits the exchange of information between differing company computer systems.

Electronic mail boxes computer terminal or computer at a work center that can receive and send messages to other terminals or computers over telephone lines.

Flow shop a type of manufacturing layout design that places equipment and personnel in accordance with the sequence of tasks required to complete a product, usually a standard product with an uninterrupted material flow. Also called mass production or a continuous manufacturing layout.

Independent demand inventory inventory items that are consumed by a customer. This inventory is also called finished goods inventory.

Inventory turnover the ratio of one year's manufacturing cost of sales (i.e., material + labor + overhead), divided by average inventory for a specific period of time (usually a year).

Manufacturing costs of sales (MCOS) material, labor, and overhead costs for a period of time.

Message standard an EDI standardized format for information used in a computer-based system.

Pallet a general purpose platform, usually consisting of wooden planking, that is used to hold a collection of items or boxes and permits them to be easily lifted by forklift trucks or pallet movers.

Produce-to-order a production schedule based on known customer demand.

Produce-to-stock a production schedule based on forecast demand and inventory storage capacity restrictions.

Order point a set inventory level in units that designate when the next order in a fixed order quantity system, like EOQ, should be placed.

Reusable containers a reusable plastic, wood, or metal container used in shipping inventory.

Scoring method a quantity method that uses evaluative scores or points to assess the comparative significance of items being compared. Also called a "scoring model."

Split delivery a method by which a large quantity is split into smaller lot sizes and spread over time to minimize material flow.

Translation software EDI computer software used to interpret messages between different computer systems.

Value-added a product's perceived utility by the customers who will consume it.

Work center a specific production facility, usually consisting of one or more workers, robots, or machines. An area where workers do their work, store their tools, and inventory is consumed. Can also be called a work station.

Discussion Questions

1. How are the JIT principles of unitary production and zero inventory related?

2. How do each of the six JIT inventory management principles improve an operation?

3. What are the differences between a JIT operation and a large-lot EOQ operation?

4. Why is the partnership between a JIT purchaser and a supplier so necessary?

5. Does the implementation of JIT principles reduce an operation's lot size? Does it ever cause an increase in the lot size?

6. What are the characteristics of JIT purchasing?

7. What is an **EDI** and how does it help a JIT operation?

8. Is it possible to operate under an EOQ system and still use JIT principles?

9. How are the assumptions used in the EOQ model the same as those in the JIT/EOQ model? How are they different?

10. What type of criteria are useful in evaluating JIT vendor performance?

11. How does the scoring method used for evaluating a JIT vendor work?

12. What can we use to evaluate the success of implementing the JIT inventory management principles? How does it work?

Problems

1. (This problem requires calculus and background on the EOQ modeling procedure.) The model in Equation 2–3 was derived from a case of only two deliveries (i.e., $n = 2$) being made. The total annual cost formula for the two delivery situation can be expressed by:

$$T_2 = \frac{CQ}{4} + \frac{QD}{Q}$$

What is the resulting JIT/EOQ order quantity model based on this total annual cost function? What is the percentage of cost savings in total annual cost for this delivery system?

2. Take the JIT/EOQ Formula 7 in Equation 2–7 and convert it algebraically to solve for the optimal order quantity Q^*. Is this resulting formula the same as that given in Formula 3? Give a numerical example.

3. Take the JIT/EOQ Formula 8 in Equation 2–8 and convert it algebraically to solve for the optimal order quantity Q^*. Is this resulting formula the same as that given in Formula 3? Give a numerical example.

4. An organization is currently using a large-lot EOQ system for ordering units of a single item inventory component. Their product is sold on a contractual basis and yearly demand is known to be 50,000 units, carrying a cost per unit of $2, and an order costs $50 to place. The firm decides that they want to shift their operation to a JIT system. What is the optimal EOQ order quantity and total annual cost? What would optimal JIT/EOQ order quantity be if the firm has chosen to reduce the lot size by splitting the order quantity into two deliveries (i.e., $n = 2$)? What would be the optimal JIT/EOQ order quantity if the firm has a maximum inventory storage limit of only 750 units?

5. An organization is currently using a large-lot EOQ system. Their product is sold on a contractual basis and yearly demand is known to be 100,000 units, carrying cost per unit is $4.50, and order cost is $76. The firm decides that they want to shift their operation to a JIT system. What is the optimal EOQ order quantity and total annual cost? What would the optimal JIT/EOQ order quantity be if the firm chose to reduce the lot size by splitting the order quantity into five deliveries? How many deliveries would they have to make and what would the optimal JIT/EOQ order quantity be if the firm has a maximum inventory storage limit of only 750 units?

6. An organization is currently using an EOQ system. Its product's annual demand is known to be 250,000 units, carrying cost per unit is $8, and order cost is $100. The firm decides that it wants to shift the operation to a JIT system. What is the optimal EOQ order quantity and total annual cost? What would the optimal JIT/EOQ order quantity be if the firm chose to reduce the lot size by splitting the order quantity into ten deliveries? How many deliveries would they have to make and what would the optimal JIT/EOQ order quantity be if the firm has a maximum inventory storage limit of only 2,000 units?

7. An organization is currently using an EOQ system. Its product's annual demand is known to be 500,000 units, carrying cost per unit is $9 and order cost is $120. The firm decides that it wants to shift the operation to a JIT system. What is the optimal EOQ order quantity and total annual cost? What would the optimal JIT/EOQ order quantity be if the firm chose to reduce the lot size by splitting the order quantity into ten deliveries? How many deliveries would they have to make and what would the optimal JIT/EOQ order quantity be if the firm has a maximum inventory storage limit of only 3,500 units?

What would the optimal JIT/EOQ order quantity be if the firm set an average inventory target of having no more than 1,000 units on hand? What would the optimal JIT/EOQ order quantity be if the firm prespecified a desire to achieve a 50 percent total cost savings from the original EOQ's total annual cost?

8. An organization is currently using an EOQ system. Its product's demand is known to be 750,000 units, carrying cost per unit is $10, and order cost is $200. The firm decides that it wants to shift the operation to a JIT system. What is the optimal EOQ order quantity and total annual cost? What would the optimal JIT/EOQ order quantity be if the firm chose to reduce the lot size by splitting the order quantity into ten deliveries? How many deliveries would they have to make and what would the optimal JIT/EOQ order quantity be if the firm set a maximum inventory storage limit of only 5,000 units? What would the optimal JIT/EOQ order quantity be if the firm set an average inventory target of having no more than 2,000 units on hand? What would the optimal JIT/EOQ order quantity be if the firm prespecified a desire to achieve a 70 percent total cost savings from the original EOQ's total annual cost?

9. A JIT operation is going to evaluate its vendors' performance using the scoring method discussed in this chapter. The three vendors A, B, and C have been scored by the operation's purchasing manager. The scores for each of the five selection criteria chosen for the analysis and for each of the three vendors, respectively, are: Price = 1, 4, and 3; On-time delivery = 3, 5, and 1; Flexibility = 1, 2, and 1; Responsiveness = 5, 7, and 1; Quality = 3, 5, and 4. Assume a score of 1 means the vendor has done a great job. What are the resulting rankings of these three vendors? Which is the best vendor? Which is the worst vendor?

10. A JIT operation is going to evaluate its vendors' performance using the scoring method discussed in this chapter. The five vendors (i.e., A, B, C, D, and E) have been scored by the operation's purchasing manager. The scores for each of the six selection criteria chosen for this analysis and for the five vendors, respectively, are: Size = 1, 4, 3, 5, and 2; Price = 4, 1, 5, 3, and 3; On-time delivery = 1, 2, 1, 3, and 1; Flexibility = 7, 6, 4, 8, and 1; Responsiveness = 2, 7, 9, 9, and 1; Quality = 9, 5, 1, 2, and 4. Assume a score of 1 means the vendor has done a great job. What are the resulting rankings of these five vendors? Which is the best vendor? If you had to drop a vendor, which should it be based on your rankings? If someone were going to challenge the results of your findings, where would they start (i.e., where is the weakness of the scoring method)?

11. A JIT operation has a yearly MCOS of $300,000 with an average inventory level of 100,000 units. In the following year their MCOS increases to $450,000 on an average inventory level of 150,000 units. The average cost per unit is $100. Is inventory utilization improving in this operation?

12. A JIT operation has a yearly MCOS of $780,000 with an average inventory level of 110,000 units. In the following year their MCOS increases to $1,000,000 on an average inventory level of 500,000 units. The average cost per unit is $200. Is inventory utilization improving?

13. Suppose a manufacturing company operating under an EOQ system has an average inventory level in the first month of 4,000 units. The accounting department reports that their cost of sales for one month amounted to $200,000 for labor, $100,000 for material, and $25,000 for overhead. During the following year, the company implements a JIT type of operation. The first month's MCOS is $300,000 under the JIT system with an average inventory level of 1,400 units. The average cost per unit is $150. What is the MCOS for the EOQ month's operation? What is the inventory turnover under both operations? Has the change in systems improved inventory utilization?

14. Suppose a manufacturing company operating under an EOQ system has an average inventory level in the first quarter of 6,000 units. The accounting department reports that their cost of sales for one quarter amounted to $1,200,000 for labor, $500,000 for material, and $80,000 for overhead. During the following year, the company implements a JIT type of operation. The first quarter's MCOS results in $2,000,000 under the JIT system with an average inventory level of 8,000 units. The average cost per unit is $50. What is the MCOS for the EOQ quarterly operation? What is the inventory turnover under both operations? Has the change in systems improved inventory utilization?

References

[1] M. A. Vonderembse and G. P. White, *Operations Management: Concepts, Methods and Strategies*, West Publishing Co., St. Paul, MN, 1988.

[2] W. J. Stevenson, *Production/Operations Management*, 3rd ed., Irwin, Homewood, IL, 1990.

[3] J. R. Evans, D. R. Anderson, D. J. Sweeney and T. A. Williams, *Applied Production and Operations Management*, West Publishing Co., St. Paul, MN, 1990.

[4] H. H. Jordan, "Inventory Management in the JIT Age," *Production and Inventory Management Journal*, Vol. 29, No. 3, 1988, pp. 57–59.

[5] R. J. Schonberger and M. J. Schniederjans, "Reinventing Inventory Control," *Interfaces*, Vol. 14, No. 3, 1983, pp. 76–83.

[6] Y. P. Gupta, "A Feasibility Study of JIT Purchasing Implementation in a Manufacturing Facility," *International Journal of Operations and Production Management*, Vol. 10, No. 4, 1990, pp. 31–41.

[7] R. G. Newman, "The Buyer-Supplier Relationship Under Just-in-Time," *Production and Inventory Management Journal*, Vol. 29, No. 3, 1988, pp. 45–50.

[8] E. A. Raether, "Issues in EDI Implementation," *Production & Inventory Management Review with APICS News*, Vol. 10, No. 9, September 1990, pp. 47-54.

[9]   A. C. Pan and C. J. Liao, "An Inventory Model Under Just-in-Time Purchasing Agreements," *Production and Inventory Management Journal*, Vol. 30, No. 1, 1989, pp. 49–52.

[10]   N. Gaither, *Production and Operations Management*, 3rd ed., The Dryden Press, Chicago, IL, 1987.

[11]   L. C. Giunipero, "Motivating and Monitoring JIT Supplier Performance," *Journal of Purchasing and Materials Management,* Vol. 26, No. 3, 1990, pp. 19–24.

[12]   J. Shaughnessy, "Tracking Vendor Performance Along with Inventory," *Automation*, Vol. 37, No. 6, 1990, pp. 52–53.

[13]   T. H. Willis and C. R. Huston, "Vendor Requirements and Evaluation in a Just-In-Time Environment," *International Journal of Operations and Production Management*, Vol. 10, No. 4, 1990, pp. 41–50.

[14]   E. J. Hay, "Implementing JIT Purchasing," *Production & Inventory Management Review with APICS News*, Vol. 10, No. 1, 1990, pp. 30–32 (Also in Nos. 2, 3, and 4).

CASE 2-1

JIT'ing to the Tune of EOQ

The Wavey-Wack Corporation (WWC) of Boston, Massachusetts, has been a long-term manufacturer/supplier of a product called "swabes" to the U.S. Navy. Their swabes product is a complex unit requiring a number of component parts that WWC has to order from countless suppliers all over the country. The lead times required to obtain the component parts and to manufacture the swabes product are quite extensive. In the past, the Navy allowed WWC to plan their deliveries around WWC's production lead times. Unfortunately, the Navy recently decided to ask for smaller and more frequent deliveries that would base production on their customer demands. It turns out that many of the other Navy suppliers were switching to a JIT operation and were now sending smaller and more frequent orders to the Navy. The Navy decided that they could reduce their inventory costs if all of their suppliers would start sending smaller and more frequent lot size orders at or around the same time. The Navy started putting pressure on WWC to implement the smaller and more frequent lot size policy request as soon as possible. WWC's business with the Navy represented over 60 percent of WWC's total sales. They had no choice but to implement the Navy smaller lot size order request.

The Navy realized that WWC had for years operated under a very large-lot EOQ system. The Navy had encouraged WWC to adopt this type of system and felt it had benefited in the past from the EOQ purchasing. The Navy didn't want to force WWC to immediately drop the EOQ order quantity system completely, but only modify it with smaller and more frequent deliveries. To make specific targeted delivery goals clear to WWC, the Navy requested that WWC use the delivery requirements in Exhibit 2-1 as targeted goals. As presented in Exhibit 2-1, the Navy sought to achieve an increase in the current 20 deliveries per year level of 1991 to an eventual 48-delivery level in 1994. This would permit the Navy to have an almost weekly delivery rate that would reduce the average inventory by more than half of its current level. Realizing the cost savings for the Navy, they added an inducement to motivate their suppliers to adopt the JIT ordering principles. The Navy signed four-year (which can be considered long-term) contracts with the suppliers. They not only agreed to continue to pay at the same cost level as the present 1991 rate, but also agreed to return to the suppliers any cost savings the suppliers could demonstrate that the Navy had achieved over the current EOQ ordering system. In other words, any cost savings WWC could demonstrate that they are providing the Navy would be given to WWC during the next four years.

EXHIBIT 2-1 • U.S. Navy Delivery Goals Program for WWC

Year	Order Quantity	Delivery Quantity	No. of Deliveries per Year
1991	800 units	800 units	20
1992	?	?	24
1993	?	?	36
1994	?	?	48

To begin the JIT-based implementation, WWC turned to their suppliers and secured long-term contracts with them for smaller and more frequent shipments of inventory. Quite a few of WWC's vendors had to be cut, but those that remained were geographically closer to the WWC operation in Boston. This closeness, it was felt by WWC management, would help to ensure a more reliable delivery rate. Some of WWC's suppliers were already using JIT and were able to implement the new ordering system quickly.

As a supplier to the Navy, WWC had to look at the prior EOQ ordering policies of the Navy to see if cost reductions would be incurred by switching to the JIT system. (These cost reductions would actually be paid to WWC since the Navy agreed to offer them as a reward for working with them in the changeover to JIT.) To begin the move to a more JIT-oriented operation, WWC re-examined the Navy's current EOQ formula calculations. They knew that the current cost they charged the Navy to place an order was $100, and the Navy's Cost Department assessed the cost of carrying a unit in stock at $5. The annual demand by the Navy for the swabes product would remain fixed during the next four years at 16,000 units per year. This results in the current 1991 EOQ optimal order (and delivery) quantity of 800 units presented in Exhibit 2-1. (We will assume in this case that the basic EOQ model presented in this chapter applies for this inventory item.) With a delivery quantity of 800 units, it will take 20 deliveries (i.e., 16,000/800) each year to satisfy the annual demand of 16,000 units.

Case Questions

1. Using the JIT/EOQ formulas for delivery splitting, what is the optimal order quantity if WWC wants to comply with the Navy's 24-delivery goal? What is the delivery quantity? What is the savings for the Navy for converting WWC to a JIT delivery system?

2. Using the JIT/EOQ formulas for delivery splitting, what is the optimal order quantity if WWC wants to comply with the 36-delivery goal? What is the delivery quantity? What is the savings for the Navy for converting WWC to a JIT type of delivery system?

3. Using the JIT/EOQ formulas for delivery splitting, what is the optimal order quantity if WWC wants to comply with the 48-delivery goal? What is the delivery quantity? What is the savings for the Navy for converting WWC to a JIT delivery system?

CASE 2-2

To Vend or Not to Vend, This is the Question

The Cuthroat Distribution Company (CDC) of Flesome, Florida, had just been notified that their major customer, Gunuser Inc., was moving to a JIT operation. Gunuser represents about 85 percent of CDC's total sales. Gunuser wanted its suppliers, of which CDC was one, to provide daily deliveries of all of the many inventory items they normally purchased from the vendors. A specific quantity and schedule were provided to CDC and the other vendors for each inventory item Gunuser purchases. Gunuser wanted an immediate changeover to the new, smaller, and more frequent delivery system of a JIT operation. Vendors were given two weeks to respond with an acceptance and a new commitment to the stated delivery schedule.

CDC is a distributor of over 300 individual items to over 500 manufacturers. About 100 different items are shipped to Gunuser every year in lot size deliveries of 10 to 10,000 units each. CDC has for several years been trying to broaden its customer base to diminish the importance of doing business with Gunuser. The move of Gunuser to a JIT operation was expected, but the rapid changeover required of suppliers took everyone by surprise. CDC had no choice but to change over or go out of business.

CDC's management decided that if they could make the changeover quickly enough, they might be able to actually increase sales with Gunuser by taking up the slack of other vendors who chose to decline the Gunuser mandate to go JIT. For a distributing organization like CDC the key to changing over was simply a matter of changing its own vendors and suppliers over to a JIT delivery system. They contacted the dozen or so vendors from when they obtained inventory for Gunuser and offered a new, long-term delivery schedule contract consistent with the stated JIT demands. Only five of the vendors (A, B, C, D, and E) were willing to commit to the new JIT-based schedule. Since any one of the five could handle the entire demand of the new schedule, CDC had to decide which of the five would be awarded the new, long-term contract.

To choose one of the five possible vendors, top management was brought into the decision process. The CDC president, the vice president of operations, and the purchasing manager all wanted a say in the evaluation of this single and most critical source of inventory supplies. It was collectively decided that each of the three would choose selection criteria and score each of the vendors on how well

EXHIBIT 2-2 • *Cuthroat Management Vendor Selection Evaluations*

President's Evaluation:

	Vendor Scores				
	A	B	C	D	E
Stability	4	3	4	8	9
Capital investment	2	6	1	5	6
Size	1	3	8	8	2
Order taking system	7	8	4	5	9
Flexibility	7	3	10	8	5
Research and development	2	3	1	1	7

Vice President of Operation's Evaluation:

	Vendor Scores				
	A	B	C	D	E
Terms of sale	5	3	4	4	9
Capital investment	4	4	6	2	1
Size	2	2	2	2	2
Order taking system	5	6	3	4	7
On-time delivery	5	6	10	1	4
Flexibility	8	10	8	10	5
Responsiveness	6	6	4	9	9
Product quality	4	1	4	8	9
Research and development	1	2	6	2	9

Purchasing Manager's Evaluation:

	Vendor Scores				
	A	B	C	D	E
Prices	4	6	6	1	1

Continued

EXHIBIT 2-2 *Continued*

Terms of sale	6	3	4	4	9
Capital investment	1	6	1	5	6
Size	2	1	8	8	2
Order taking system	6	8	4	5	9
On-time delivery	4	6	8	1	4
Exact quantity accuracy	5	3	4	9	3
Flexibility	3	3	9	8	5
Responsiveness	7	1	1	6	7
Product quality	3	1	4	8	9

they would be able to serve CDC operating under a JIT system. Data from all five of the vendors was quite extensive and the importance of the suppliers to the CDC operation had made a close relationship with each of them a requirement in past business activities. All three of the evaluators knew and could fairly assess the scores for each of the five vendors. These selection criteria and scores are presented in Exhibit 2-2. A score of 1 in Exhibit 2-2 means that a vendor is judged as probably being the best at providing that particular service to CDC. A score of 10 represents an evaluation of probably the worst service. While many of the selection criteria were the same for each evaluator, upper level management tended to select items that related to the size and stability of the vendor, whereas the purchasing manager selected criteria more concerned with day-to-day items like the accuracy of quantities in a delivery.

Case Questions

1. If all of the evaluation scores are to be used in the evaluation of the vendors, how can the scores from the three evaluators be used to make the selection decision?

2. Using the approach stated in Question 1, which vendor should CDC choose? Support your decision.

3. Are there any limitations on the selection decision made in Question 2? If so, state them. If not, explain why not.

CHAPTER THREE

JIT Production Planning and Scheduling

Chapter Outline

Learning Objectives

After completing this chapter, you should be able to:

1. Explain what JIT production management principles are and how they can improve a manufacturing operation.
2. Explain how a JIT facility layout is different from a large-lot facility layout.
3. Describe how a synchronized pull system is used in a JIT operation.
4. Suggest tactics for reducing setup costs and improving communication in a JIT operation.
5. Explain how the mixed model scheduling method is used to determine a JIT production schedule.
6. Explain how single card and dual card *kanban* systems work.
7. Describe a new technology used to support a *kanban* card system.
8. Describe several management performance measures used to evaluate a JIT operation.

Introduction

A **production plan** is a general guide that defines the overall level of production an operation expects to produce over an extended period of time such as a month, a quarter, or a year. It usually defines the number of units or the production rates of the various products an operation will produce to meet customer demand. As a production plan is divided into units of time, it becomes a **production schedule,** a detailed plan defining the exact number of units that will be produced per hour, day, or week [1,2,3,4].

In large-lot operations, production planning usually begins with an examination of intermediate-term customer demand forecasts of one to three years. Based on these forecasts management can plan aggregate labor, aggregate inventory, aggregate production rates, and aggregate capacity needs. Referred to as **aggregate planning,** these aggregate labor, production, inventory, and capacity needs are then balanced to achieve the desired production rates necessary to meet customer demand. Once this is accomplished, the production activity is divided into smaller time periods to define a specific unit production plan called a **master production schedule (MPS).** (It is assumed students can prepare an MPS.)

In a JIT operation, customer demand determines the MPS. A JIT operation simply has to be able to shift its production rate on a weekly or daily basis to match the shifting demand of the operation's customers. As we discussed in Chapter 2, long-term supplier contract ordering quantity boundaries have to be set to allow for some flexibility in purchasing inventory. Production planners must design and equip their operations to be able to shift production rates within those stated boundaries in such a way that the waste of labor, materials, and equipment is minimized. A JIT operation not only has to be able to shift production rates, but also be able to shift production activity from one model of a product to another model quickly and with a minimum of setup costs.

To help JIT managers accomplish their production planning and scheduling, a number of principles and methodologies have been developed. The purpose of this chapter is to provide a basic understanding of JIT production planning and scheduling principles. In addition, this chapter describes two specific scheduling methods used to support JIT operations: the **mixed model scheduling method** and *kanban* systems. To help users identify the success of JIT production planning and scheduling, a series of JIT measurement formulas are also presented.

JIT Production Management Principles

There are many production management policies, rules, and procedures that are related to production planning and scheduling in a JIT operation. Several of these can be characterized as JIT production management principles. In this chapter we seek to expand on the general principles presented in Chapter 1 that are focused directly on production planning and scheduling. The related production planning and scheduling principles include:

1. Seek uniform daily production scheduling.
2. Seek production scheduling flexibility.
3. Seek a synchronized pull system.
4. Use automation where practical.
5. Seek a focused factory.
6. Seek improved flexibility in workers.
7. Cut production lot size and setup costs.
8. Allow workers to determine production flow.
9. Improve communication and visual control.

1. Seek Uniform Daily Production Scheduling A **uniform daily production schedule** is a day-to-day schedule of production where there is little or no variation in production quantities between days. To accomplish a uniform daily production schedule requires the planning activity called **load leveling.** Load leveling is a production plan established to allow unitary levels of each product to have flexibility to change from month-to-month, but remain the same each day during a monthly planning period. Product quantity changes are permitted on a monthly basis to meet changes in customer demand, but production during each day of the month is held level.

In Figure 3-1 a monthly and daily scheduling situation is presented. In Figure 3-1(a) a monthly schedule for three products (*A*, *B*, and *C*) is presented as it might be scheduled in a large-lot operation. The amount of production of product *A* each day during the large-lot operation may vary because of equipment or personnel limitations for this particular product. Indeed, if product *A* required unique equipment, the operation would have to make a considerable investment in equipment to keep all the rest of their personnel busy during the long run time of product *A*.

Alternatively, the monthly production schedule could be divided into daily production schedules, each having the same production level. In this way, small-lot production of all three products can be scheduled each day, and at the end of the month will equal the total month's required production. By requiring small-lot production from all the **work centers** in an operation each day, this scheduling principle will minimize idle capacity for both workers and equipment.

One of the scheduling methods used to achieve a uniform daily production schedule is called a *mixed model schedule*. This method will be discussed in the next section of this chapter.

FIGURE 3-1 • *Monthly and Uniform Daily Production Schedules*

(a) Monthly Production Schedule for Three Large Lots

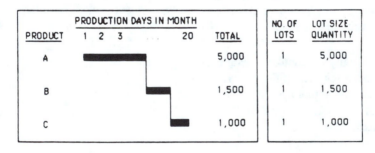

(b) Uniform Daily Production Schedule for Sixty Small Lots

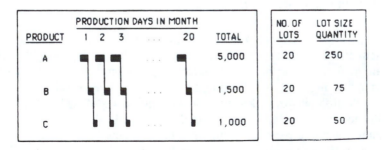

2. Seek Production Scheduling Flexibility Production capacity can be defined as the output capability of a work center.

From prior JIT principles we know that a production schedule is set to meet known customer demand. Since the production capacity by which the demand is to be met is under the control of management, we seek to establish a level of capacity that will permit needed flexibility to meet minor shifts in customer demand. A JIT operation has to have sufficient flexibility to make daily shifts in scheduled production (and all other support systems including vendors supplying inventory) to match the actual shifts in market demand.

Unfortunately, considerable management decision-making effort is necessary to balance the cost of this flexibility. Too much idle capacity will invite waste, and too little will cause inventory shortages, line idleness, and a great number of inefficiencies. Fortunately, there are several JIT strategies that can be used to guide management decision making in this area. One of the most commonly recommended strategies used in JIT operations involves the simple effort of scheduling production at less than full capacity. How much less should be scheduled depends on the excess or shortage costs of capacity. Ideally, capacity and production should equal one another. While the focus of effort in a JIT operation is usually aimed at avoiding excess capacity (which is viewed as waste), an excess capacity strategy is offered as an introductory method to eventually be eliminated. When a JIT operation is starting up, scheduling is useful to avoid stressing workers who must learn new methods and work habits. By scheduling at less than full capacity the workers are given time to get used to JIT, and they spend some time on improving their own work activities, receiving more training, and servicing equipment. Eventually, improvements will result in a more nearly full capacity schedule. (As we will discuss in the next chapter, the time not spent on work activities should not be wasted, but it is devoted to other important JIT quality control tasks.)

3. Seek a Synchronized Pull System A **pull system** operates only in production environments where known customer orders drive the production effort. The schedule of production is pulled by, and hopefully synchronized to, the actual customer demand. Since the customer is outside the JIT operation, the customer's placing of an order acts to pull the inventory through and out of the production operation. Alternatively, in a **push system** management decides what is to be produced based on forecast demand, and seeks to push production and inventory through the operation to meet the forecasted demand. Under this JIT principle, an operation seeks a scheduling system that will synchronize production activity to the demand pulled through the operation by known customer orders.

One of the most commonly used pull system scheduling methods that can be used to support a JIT-type operation is a card system called *kanban* (pronounced kahnbahn). This scheduling method will be discussed in a later section of this chapter.

4. Use Automation Where Practical In a JIT production operation, automation usually involves **robots, electronic sensors,** and **automated handling systems.** As a general principle, JIT production managers automate only those jobs that automation can perform better than humans. In effect, this principle seeks to allocate resources on the basis of rational economics. Humans are intelligent and highly flexible in work assignments. Automated robots are highly efficient and accurate in their work assignments. In a JIT operation, human workers are assigned jobs that require greater flexibility than can be performed economically by robots. In other situations, where jobs are very physically exhausting, simple, or boring, robots are employed because they are more efficient. In some situations computer-controlled electronic sensors are far more accurate at detecting defects and flaws than humans. This is particularly true in high volume, small item inspections (e.g., milling work on a machined part) or where a high level of accuracy is required (e.g., high precision measurements).

Humans are often better in spotting flaws in large items (e.g., a flaw in a paint job on a wall in a room) or on products where sensors cannot be easily placed (e.g., on the side of a dam being built).

Automation, for automation's sake, is wasteful. Automation in general is costly and tends to limit the flexibility of an operation to the capabilities of the automated unit. During the 1980s, many organizations that adopted JIT tended to move away from automated systems. In the 1990s new automated systems are now finding linkages with JIT. We will be discussing one of these automated linkages in a later section of this chapter and throughout the rest of this book.

5. Seek a Focused Factory A **focused factory** is one that is dedicated to producing a limited number of different products with a limited number of production processes. A factory that is focused on producing one type of product using a single, unchanged set of production processes is the ultimate focused factory. In a JIT operation where flexibility is important, a focused factory is only justified where there is a sufficient and continuous demand for a single product or a **product family,** a group of products that share similar production requirements and/or inventory components. Part of the principle of a focused factory in a JIT operation is to balance the limited number of production processes with the need for flexibility in producing a number of similar products sharing those production processes in a product family. A focused JIT factory has more flexibility than the highly automated traditional focused factory of the past.

A JIT-focused factory seeks to minimize product flow and work-in-progress (WIP) lead time wastes. In Figure 2-1 in Chapter 2, the idea of minimizing material handling was illustrated by a change in plant layout.

In single product situations where sequential production processes permit a **flow shop** layout, material handling and WIP flow time can be minimized. In more complex or multi-product situations the setup time for large-lot production runs is considerable, and a departmental layout is more commonly used. In Figure 3-2(a) a departmental layout is presented. The dark lines between the departments represent the flow of materials. There are four manufacturing departments in Figure 3-2(a) that are labeled for a total of six component parts they manufacture for the two finished products (Product 1 and Product 2) of the plant. Let's say component parts 1, 2, 3, and 4 are assembled together to make a unit of product 1, and component parts 5 and 6 are assembled together for product 2.

Typically, the raw materials come into the large-lot operation in Figure 3-2(a) in the Receiving Department, and then they are stored in the Raw Materials Inventory Department. From there the various raw materials are sent to the four component manufacturing departments. Setup times between the differing component parts in Component 1 & 2 and Component 4 & 5 Departments would require substantial time and differing equipment to meet the production needs of the components. Once converted into component parts, they are stored in the Component Parts Inventory Department until required in the Assembly Department. Large lots of product 1 would require large lots of component parts 1, 2, 3, and 4, and large lots of product 2 would require large lots of component parts 5 and 6. When assembled, the finished products would be stored in the Finished Goods Inventory Department until ordered by the customer and finally shipped in large lots from the Shipping Department.

In a focused JIT operation, the required layout is substantially smaller and requires less materials handling. The layout in Figure 3-2(b) assumes that substantial demand exists to warrant a continuous dedication of a production facility for each of the two products separately (i.e., the upper portion of the facility is dedicated to product 1 and lower portion for product 2). In Figure 3-2(b) materials are brought in through separate Receiving Departments and feed directly to separate component manufacturing departments. By dedicating separate manufacturing departments to the six individual components, setup times are eliminated, and some wasteful duplications in equipment are excluded. Most of the inventory storage space is eliminated in the unitary, demand pull, uniform production environment of the JIT operation. Also, the material and WIP flow is chiefly in a linear direction through the facility, minimizing flow time.

An alternative JIT layout to the one in Figure 3-2(b) could have had the Assembly (or Receiving) Department combined into a single department. Such an assembly department would have to possess sufficient flexibility to accommodate the two differing products. To achieve this flexibility a JIT operation usually uses a philosophy of layout called **group technology (GT)**. GT is a philosophy that identifies the similarity or sameness of parts, equipment, or production processes to take advantage of common setups.

FIGURE 3-2 • *Comparison of Large-lot and JIT Layout Designs*

(a) Large Lot Operation Layout

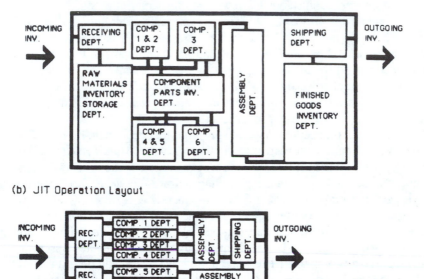

(b) JIT Operation Layout

To illustrate the GT philosophy, let's look at a simple layout situation. Suppose we have a "product family" of three VCRs (models *A, B,* and *C*) that share the same basic component parts except for a few optional components used to differentiate the different models that make up the product family. The addition of the optional components requires some differing production activities to complete each different model. In a production layout work activities are assigned to **work centers** that are arranged along a production flow line. In JIT GT layouts we usually call the production flow line a **group technology cell.** The GT cell is usually structured in the form of a U or a C. A U-shaped GT cell for this VCR illustration is presented in Figure 3-3. The idea of the GT cell is that work is started at one end of the GT cell, and after going through the various required work centers along the U, a finished unit results at the other end.

In Figure 3-3 the GT cell is set up for product *A*. The arrows in the figure represent the movement of work center tables and equipment away from or toward the line as needed for each different VCR model. This type of worker, work center, and equipment movement is a common part of a GT cell production setup effort. Inventory materials handling efforts would also be reconfigured to supply the individual work centers with the least amount of movement. Only those work centers required for producing *A* are moved toward the U line and supplied with a worker. The other work centers are temporarily moved out of the way to provide workers with additional space. The Japanese experts actually recommend that work center tables and machines be equipped with casters and rollers so they can be rolled back and forth by workers as needed [5, p. 121]. This production flexibility requires a greater flexibility in the skill of the worker. We will discuss this need in the next JIT production principle.

FIGURE 3-3 • *Group Technology Cell: Product "A" Layout*

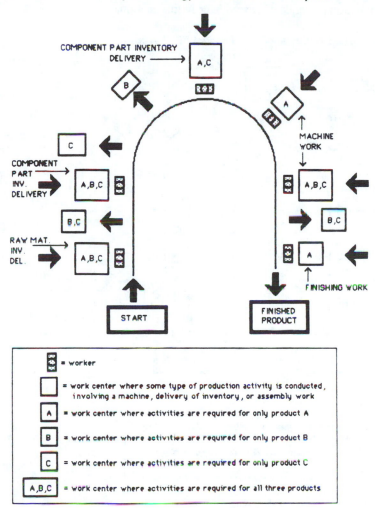

Notice in Figure 3-3 that the first two work centers (i.e., those with people stationed next to them) on the line are capable of producing all three products. Since some of the work centers are capable of producing other products, they will not have to be substantially changed or moved when a different product is ordered from the line. Notice how little the layout is changed when it is reconfigured for product *B* as presented in Figure 3-4. This will result in the production setup lead times being minimized when a change in production is required for a differing product. This helps to achieve the necessary JIT reduction in setup costs of the seventh JIT production principle discussed later in this section.

FIGURE 3-4 • *Group Technology Cell: Product "B" Layout*

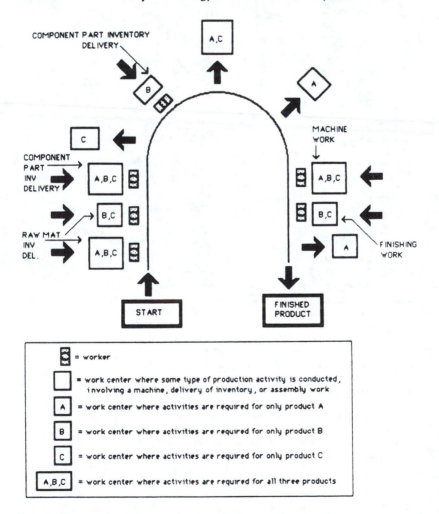

A focused factory also means a smaller and simpler type of operation. As factories grow to meet the growth in the organization's sales and need for greater diversity in the range of different products in their product families, the amount of production complexity in planning and control systems also grows. In a JIT environment, management should seek simplicity in the production methods used to produce goods and in the layout of the operation. On average, Japanese JIT manufacturing plants are smaller in size than U.S. plants, yet they produce more, and offer greater diversity and flexibility in the range of different products they can produce. This increased production productivity per square foot and flexibility has come by years of implementing JIT principles that have generated labor-, space-, and equipment-saving ideas from managers and workers. By keeping work methods simple and reducing system complexity, JIT plants avoid the need for the more complex and involved computer-based methods U.S. manufacturers typically use. Often the simple act of splitting a large manufacturing plant into two smaller and more focused plants (i.e., each producing unique products) can cause synergistic reductions (i.e., a reduction where $1/2 + 1/2 < 1$) in system complexity.

6. Seek Improved Flexibility in Workers In a JIT operation we seek to hire the most highly qualified, multi-skilled workers. Doesn't everybody want the most highly qualified and multi-skilled workers? No, not if the objective driving the operation is to just minimize cost. Most workers are paid at their highest level of skill achievement. Management logically doesn't want to pay a worker a high-skill wage for low-skill tasks. Common practice in cost minimizing organizations is to hire minimum qualified personnel to get the job done at the least cost. In these types of operations management actively encourages turnover to eliminate overly qualified personnel. The problem with this mentality is that every time a change in production is required, new workers with the minimum qualifications have to be hired and others fired. Many U.S. operations experience an employee turnover rate that is the equivalent of changing their entire manufacturing personnel every five years. This rapid turnover inhibits achievement of many of the long-term JIT principles, including seeking to eliminate waste, seeking continuous flow improvement, respect for people, and the long-term emphasis discussed in Chapter 1.

Several tactics are employed in JIT operations to overcome the problems of rapid turnover and improving worker flexibility. Two of these tactics include cross training and the use of part-time workers.

Extensive **cross training** of personnel is designed to provide job assignment flexibility. Workers should be cross trained to perform a variety of different jobs. This permits management flexibility in making job assignments when production requirements shift, and also helps to assure workers they will be retained during periods of change. The benefits of providing stable employment to workers cannot be overstated. The idea of what the Japanese used to call lifetime employment clearly has benefited Japanese firms. Unfortunately, as the Japanese workers are now learning, and as the U.S. workers have always known, there is no such thing as lifetime employment. Doesn't the broad range of skills of long-term, cross-trained employees require a higher pay scale? Yes, in general, it does increase total wages. But the savings advantages of eliminating continual training costs caused by employee turnover, costs of rescheduling production due to a shortage of skilled labor, costs of idle labor caused by an inappropriate skill that contractually must be paid, and poor product quality caused by a lack of worker skill, can easily outweigh the increase in total wages caused by having a better-trained worker. The multi-skilled JIT worker will have confidence in knowing that he or she is of greater value to the operation, and hence, have a greater sense of job security. Cross-trained workers also permit the manager to have greater flexibility in making work assignments, and hence, will not have to cause rapid turnover to minimize total costs.

Another tactic used to achieve greater worker flexibility is the use of part-time workers. Many JIT firms employ as much as half of their operations personnel as part-time workers. There are a number of good reasons why part-time workers are beneficial to an operation. Part-time workers do not always receive as many financial benefits as the full-time workers, so they tend to be less costly; they are expendable during downturns in customer demand, which allows management greater flexibility to match customer demand requirements; and their temporary presence provides full-time personnel with a greater sense of job security (because part-timers would be fired before full-timers in periods of declining customer sales). Why would anybody want to be a part-time worker under such uncertain working conditions? Part-time work permits first-time workers an opportunity to learn what work is about. In general, part-time work provides greater working hour flexibility, provides a second income for working families, and is a way to qualify for a full-time job when one opens up.

7. Cut Production Lot Size and Setup Costs The JIT inventory management principle of cutting inventory ordering lot sizes and increasing the frequency of orders can create a major production scheduling problem. As smaller orders of inventory are brought into a JIT operation, they force smaller but more frequent production runs to use the incoming inventory. The problem is in handling the increase in setup costs caused by the smaller and more frequent production runs. To achieve smaller lot size production economically requires JIT operations to reduce setup cost. This required JIT relationship between lot size and setup cost is illustrated in Figure 3-5. In the procedure called **economic manufacturing quantity (EMQ)** analysis, the cost functions in Figure 3-5(a) are mathematically used to seek a cost-minimized (i.e., minimizing total **carrying costs** and **setup costs**) lot size value of EMQ_1. (This procedure is based on the same methodology as the "EOQ" modeling approach discussed in Chapter 2.) If setup costs can be reduced as presented in Figure 3-5(b), the EMQ lot size is reduced from EMQ_1 to

EMQ_2. A setup cost reduction helps to achieve the reduced lot size production schedule necessary to support the JIT inventory principles.

How are setup costs reduced in a JIT operation? There are several tactics that can be used, and all are focused on eliminating waste. From Chapter 1 the five S's of proper arrangement *(seiri)*, orderliness *(seiton)*, cleanliness *(seiso)*, cleanup *(seiketsu)* and discipline *(shitsuke)* can help reduce setup times at work centers. A major loss of setup time can be saved by being able to obtain the right tool to do a job. If a work center is clean and the tools are properly arranged, they can be efficiently called into action when required. To help make it easy for workers to put tools back in their proper place, the image of the tool can be sketched by the tool's holding hook on a wall in the work center where it is to be used. At any time of the day, both worker and manager will be able to observe easily which tools are not hanging in their proper place and are missing.

Another simple tactic to reduce setup time is the practice of planning and staging production activities at the work centers. In the GT cell configuration, workers who are finished with one type of product can begin the **changeover** (the tasks of setting up the equipment and inventory for a production run) to another type of product as soon as the last unit of one lot has passed through their

FIGURE 3-5 • *Impact of JIT Setup Reductions on Lot Size*

(a) Basic Large Lot EMQ Cost Graph

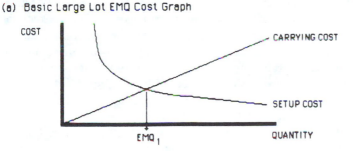

(b) Impact of Setup Cost Reduction on EMQ

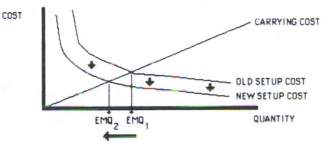

production schedules to stage the proper sequence of tools for the day's assigned tasks in an effort to help minimize setup lead times. JIT experts recommend a goal of eliminating all changeover or setup time in situations where the production processes or cells have to be completely stopped [11, p. 144]. Part of the act of planning and staging work also includes some effort in simplifying the setup requirements. While not always possible for complex production processes, work simplification involves the redesign of the setup job tasks to make the tasks easier for the workers to remember and perform. Sometimes this effort may require the specialized training of an industrial engineer who can design new tools or work center layouts that simplify setup effort. In other cases, a manager's application of the five S's might simplify and shorten setup lead times.

Finally, this setup cost reduction principle should be viewed as a goal of continual improvement. There will always be some setup costs for a production run, but seeking a goal of zero setup costs should be viewed in the same way that we seek a goal of zero inventory.

8. Allow Workers to Determine Production Flow Each work center along a GT cell or assembly line should be designed to permit the worker to decide the flow of production. In other words, the worker should decide if he or she has completed the work assignment on a product before the item is sent to the next work center. Many JIT operations design their production lines to stop WIP items at work centers until the worker approves sending it along. The worker who believes the work assignment is complete activates a brake release on the item to permit it to continue to the next work center. In this way, the workers control the flow of the line and also accept the responsibility for doing a job right—they can't say they didn't have enough time to complete their work. We will discuss the quality benefits of this worker responsibility for production control in the next chapter.

By letting workers control the line flow, management can better observe where production line imbalances or production problems exist that may affect the production schedule. As queues of work-in-process back up at work centers, management can identify where work assignments need balancing, where additional worker training may be needed, when an improvement in setup lead time is needed, or where flawed components are arriving from a vendor. Most important, this JIT principle helps to identify problems that slow the production process. Once identified, managers and workers can work together to solve the problems and improve the product flow. This production planning and problem-solving effort triggers the many benefits of the productivity cycling process discussed in Chapter 1.

How can U.S. manufacturers allow workers to determine production flow when they have to keep up with an MPS? The JIT firms are big users of **production quotas** as short-term unit production goals. Work center personnel are asked to establish reasonable daily unit production quotas as a means of achieving management MPS set goals for the entire operation. The quotas are set and well posted in a production facility to act as a motivator to exceed (or an embarrassment if not met). Either way, the quota sets daily goals that are easy to understand and act to motivate workers to keep on track with the MPS. In a JIT operation the posting of these quotas tends to be for groups or teams of workers rather than for individuals. Presenting the quota in this way builds group and team spirit, which can be a powerful force for productivity.

9. Improved Communication and Visual Control Research has shown that improved communication is very necessary for JIT success [6]. Specific areas of action or communication tactics that will facilitate JIT performance include planning meetings that involve multiple departments, joint departmental communication duties that involve all areas of a production operation, specific departmental meetings that communicate the continued ideas and importance of JIT, continual written information discussing JIT principles, encouraging informal communication between workers on the subject of JIT, and individual meetings between managers and their subordinates to discuss JIT goals.

Improved communication not only involves the discussion of JIT goals, but also seeing that those goals are being achieved. JIT operations must be designed to facilitate what is sometimes called **visibility management,** which enhances management control and correction when deviation from goals is observed [11, pp. 58-64]. The posting of group or team productivity is one example of visibility management. The posted information acts to provide easily seen and understood production control information. Visibility management also involves the entire layout design of a production facility. By designing a facility to facilitate the observation of deviation from JIT goals, managers and workers will be motivated to solve problems that may be causing inefficiencies in production more quickly. One way JIT operations can facilitate a more visible work environment is by eliminating the interior walls of a plant. This prevents workers and managers from hiding production problems. Possible product assembly problems can be avoided by placing sign boards on the walls of the plant that depict defective work. The sign board, illustrating potential problems workers commonly commit, acts as a continual reminder of what workers should be careful to avoid in the performance of their job.

Visibility management also is directed at saving worker production time. Let's look at one example of how improving visibility in a control mechanism can save worker production time. Suppose a worker had to check and make sure a series of four gauges are at optimum pressure levels in order to initiate a production process. Suppose each gauge measures a different part of the production process, and as such, has a different optimum level of pressure. Even if the gauges were actually at their optimal levels as they are in Figure 3-6(a), the worker would have to know or take the time to look up the optimum levels of each gauge to be sure they are all at the desired pressure level. Now suppose we take the same gauges and reconfigure them as in Figure 3-6(b). In this configuration the optimal pressure level for each gauge will be achieved when the gauge's arrow pointer is in a vertical direction pointing up. Now the worker will not have to know what the individual optimal levels are to determine if the pressure is off, only that a deviation exists if the arrow is not in the vertical position. Using this improved visual adjustment to the gauges, a worker can more quickly determine optimal pressure levels, saving worker production time and improving worker productivity. The JIT production planning and scheduling principles presented in this section are in no way complete. Those principles presented provide only a brief introduction to the basics of JIT production management. They are an attempt to illustrate the simplicity and beneficial logic in JIT production management.

A JIT production facility doesn't require great investments in technology, but must have the full and continual participation of workers and management. In addition to the guiding principles, there are a number of methodologies that support JIT production planning and scheduling. One of these JIT methodologies is called the mixed model scheduling method.

FIGURE 3-6 • *Visibility Management Gauge Example*

(a) Typical Gauge Arrangement

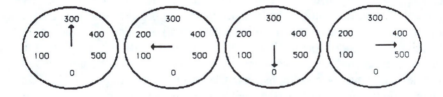

(b) Improved Visibility Management Arrangement

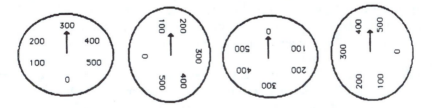

Mixed Model Scheduling Method

Suppose a company has just set up its first GT cell. They want to change their monthly large-lot schedule to a JIT daily uniform schedule. This type of situation was previously presented in Figure 3-1. The company would begin by taking the monthly product mix demand in Figure 3-1(a) and converting it into the 20 days of uniform demand as in Figure 3-1(b). The mixture of products (or different product models) produced in a day requires mixed model scheduling. The question becomes one of determining the size and number of lot runs in each day for each of the products in the mix. The mixed model schedule in Figure 3-1(b) appears to suggest three runs of 250 units of product A, 75 units of product B and 50 units of product C accomplishes the uniform daily production schedule, which it does. But if smaller daily production runs improve production efficiency, wouldn't a cut still further to 6, 12, or 90 smaller production runs within a period of a day improve efficiency? The answer is yes; smaller, almost unitary production runs will create the smoothest possible production flow. The smoother the production flow, the more productivity is maximized by eliminating production line imbalances. Also, the smoother the production flow, the less the volatility of demands on inventory planning, thus maximizing the JIT inventory management benefits.

If smaller production runs per day are beneficial, how do we go about scheduling production runs? Specifically, how should the differing product models be sequenced throughout the day's production schedule? The answer to this scheduling problem is found by using the mixed model scheduling method. The **mixed model scheduling method** is a procedure that can be used to determine the minimum number of units to sequence in a production run for a daily production schedule. This method is based on minimizing lot size and determines the size of production runs. While this method seeks to achieve the JIT principle of unitary production, lot sizes for products might have to be greater than one to maximize production flow.

One approach to the mixed model scheduling method consists of the following steps:

1. Determine the Daily Unitary Production Schedule for the Mixture of Product Models   Similar to Figure 3-1, the monthly demand for each product that is to be produced during the month is divided by the number of working days in that month. The ratio for each product determines the number of units that must be produced each day of the month to meet the MPS goal for the month.

2. Determine the Cycle Time for Each Product   The **cycle time** for a product is the amount of time required between successive completions of a product, or the amount of time to complete one unit. In other words, if it takes 60 minutes to complete 30 units, the cycle time for each unit is two minutes (i.e.,

60/30). For illustrative purposes, let's say the cycle time for product *A* is three minutes, product *B* is six minutes, and product *C* is six minutes.

3. Determine the Reciprocals of the Cycle Times for Each Product In the case of this illustration they would be, respectively:

$$\frac{1}{\text{cycle time for product A}}, \quad \frac{1}{\text{cycle time for product B}}, \quad \frac{1}{\text{cycle time for productC}}$$

or

$$\frac{1 \text{ unit}}{3 \text{ minutes}}, \quad \frac{1 \text{ unit}}{6 \text{ minutes}}, \quad \frac{1 \text{ unit}}{6 \text{ minutes}}.$$

4. Determine the Ratio of the Total Minimum Number of Units in a Sequence to the Sequence Time This is accomplished by first simplifying the denominator of the ratios in Step 3 into a common number and adding the fractions together to give the desired ratio. This is just a simple mathematical conversion that determines the **total sequence time**, the total amount of cycle time required to complete one sequence of production. A common number for all the denominators is 6 minutes, which gives the following ratios for each product, respectively:

$$\frac{2 \text{ units}}{6 \text{ minutes}}, \quad \frac{1 \text{ unit}}{6 \text{ minutes}}, \quad \frac{1 \text{ unit}}{6 \text{ minutes}}.$$

Adding these ratios together gives the desired ratio:

$$\frac{\text{Minimum Units in Sequence}}{\text{Total Sequence Time}} = \frac{2 \text{ units}}{6 \text{ minutes}} + \frac{1 \text{ unit}}{6 \text{ minutes}} + \frac{1 \text{ unit}}{6 \text{ minutes}}$$

$$= \frac{4 \text{ units}}{16 \text{ minutes}}$$

The interpretation of this ratio determines the production run size of a single, repeated sequence of units to be produced. In the sequence above the minimum number of units to have in a sequence is four—we need to produce two units of *A*, one unit of *B*, and one unit of *C*. The total time to complete this sequence of four units of production will be 18 minutes. This sequence of four units produced would be repeated 26.67 times in an eight-hour day: (8 hours x 60 minutes)/18 minutes).

What would the ratio be if odd-numbered cycle times were used, like 2 minutes, 3 minutes, and 5 minutes, respectively, for the three products?

Repeating the steps above, we have:

$$\frac{1 \text{ unit}}{2 \text{ minutes}}, \quad \frac{1 \text{ unit}}{3 \text{ minutes}}, \quad \frac{1 \text{ unit}}{5 \text{ minutes}},$$

which are converted into:

$$\frac{15 \text{ units}}{30 \text{ minutes}}, \quad \frac{10 \text{ units}}{30 \text{ minutes}}, \quad \frac{6 \text{ units}}{30 \text{ minutes}},$$

and finally,

$$\frac{\text{Minimum Units in Sequence}}{\text{Total Sequence Time}} = \frac{15 \text{ units}}{30 \text{ minutes}} + \frac{10 \text{ units}}{30 \text{ minutes}} + \frac{6 \text{ units}}{30 \text{ minutes}}$$

$$= \frac{31 \text{ units}}{90 \text{ minutes}}.$$

This ratio reveals the minimum sequence consists of 31 units (15 units of A, 10 units of B, and 6 units of C). The total time to complete this sequence of 31 units of production is 90 minutes. This sequence of 31 units produced would be repeated 5.67 times in an 8-hour day: (8 hours x 60 minutes)/90 minutes)

5. Determine the Unit Order Sequence Schedule   Scheduling the order in which the individual products will be produced in the repeatable sequence should be based on the JIT principle of unitary production wherever practical. That is, the order of the product sequence should be set to achieve unitary production unless practical production constraints require a larger lot size. Some manufacturing equipment with fixed setup times might constrain lot size by forcing a fixed number of units to be produced at one time. For example, if a machine has a fixed setup time and is designed to work on six units at one time, it would be impractical to use a smaller lot size than six, since it would only increase the number of setups. In this situation the sequence of products should be set in an order that matches six units of the one product together in a single lot.

If we don't have limiting constraints, we can aspire to achieve the JIT unitary principle. The objective is to mix production so no two units of the same product are next to one another. In the example above where we arrived at a sequence of 4 units in 18 minutes, we could arrange the sequence in one of any of the following combinations:

Sequence

No.	1	2	3	4
1	A	B	A	C
2	A	C	A	B
3	A	B	C	A
4	A	C	B	A
5	B	A	C	A
6	C	A	B	A

There is no easy statistical formula that is worth the effort to use to find all of the possible combinations of production sequences. We simply use a little judgmental effort to sequence the products in an order that will minimize the repetition of the same product in the job sequence. Once the single sequence above is chosen, it is repeated throughout the day until the daily production quota is achieved. Since we never produce the same product twice in a row, we have again achieved a unitary production schedule for this product mix.

This scheduling of products logically assumes that if there is sufficient capacity to achieve the production quota under the large-lot operation, its redistribution into smaller lot sizes will still achieve the quotas. If there is not sufficient production capacity, the flexibility under the JIT operation permits a more rapid response than is possible under the usual large-lot production planning system.

In the other example above where we arrived at a sequence of 31 units in 90 minutes, we might start the sequencing effort by looking at the problem as one of placing production based on the proportion of units each product represents in the total sequence. In this problem we have 31 positions into which to put units of product:

$$\frac{\text{Product} \ \_ \ \_ \ \_ \ \cdots \ \_}{\text{Sequence} \ 1 \ 2 \ 3 \quad 31}$$

Since 15 of the 31 positions are units of A, we know that about half of the units in the sequence will be allocated for product A. Logically, this requires every other position to be an A position to keep A from being placed next to itself in the sequence.

$$\frac{\text{Product} = A \ \_ \ A \ \_ \ A \ \_ \ A \ \_ \ A \ \_ \ A \ \_ \ A \ \_ \ A \ \_}{\text{Sequence} = 1 \ 2 \ 3 \ 4 \ 5 \ 6 \ 7 \ 8 \ 9 \ 10 \ 11 \ 12 \ 13 \ 14 \ 15 \ 16}$$

$$\frac{\text{Product} = A \ \_ \ A \ \_ \ A \ \_ \ A \ \_ \ A \ \_ \ A \ \_ \ A \ \_ \ \_}{\text{Sequence} = 17 \ 18 \ 19 \ 20 \ 21 \ 22 \ 23 \ 24 \ 25 \ 26 \ 27 \ 28 \ 29 \ 30 \ 31}$$

If we had more than half As, let's say two thirds As, then we would have to place two As next to one another. If we had three quarters As, then we would have to place three As next to one another in the sequence, and so on. This judgmental guideline acts as a tactical approach to determining lot size in the JIT small-lot operation.

Now we might look to the next most common product, ten units of B. Seeking to avoid the pairing of product B we can simply spread the Bs evenly over the open positions in the unfilled portion of the sequence as follows:

$$\frac{\text{Product} = A \ B \ A \ B \ A \ B \ A \ \_ \ A \ B \ A \ \_ \ A \ B \ A \ \_}{\text{Sequence} = 1 \ 2 \ 3 \ 4 \ 5 \ 6 \ 7 \ 8 \ 9 \ 10 \ 11 \ 12 \ 13 \ 14 \ 15 \ 16}$$

Product = A B A A B A A B A B A B
Sequence = 17 18 19 20 21 22 23 24 25 26 27 28 29 30 31

The final allocation of the six *C*s gives us the sequence of scheduled unit production:

Product = A B A B A B A C A B A C A B A C
Sequence = 1 2 3 4 5 6 7 8 9 10 11 12 13 14 15 16

Product = A B A C A B A C A B A B A B C
Sequence = 17 18 19 20 21 22 23 24 25 26 27 28 29 30 31

The total time to complete this sequence of 31 units of production is 90 minutes, and it would be repeated until the daily production quota is achieved. Since no two of the same product are next to one another, we have again achieved a unitary production schedule for this product mix.

The mixed model scheduling method presented above is just one of many JIT methods useful in implementing a unitary production schedule. In the next section we will examine another JIT scheduling method based on paper cards called *kanbans*.

Kanban Card Systems

A *kanban* is a production scheduling and inventory control card system [7]. The Japanese term *kanban* can be interpreted as meaning "card." A kanban system uses paper cards to control the scheduling of production activities and the use of inventory. The kanban cards may be disposable 4 by 8 inch cards or plastic reusable cards. While a JIT system does not have to have a kanban system to operate, a kanban system supports a JIT environment in that it is capable of implementing unitary or small-lot production.

There are several types of kanban cards that are each used to signal the authorization of some production or inventory activity. These kanban cards include a production authorization card, a vendor authorization card, and a conveyance authorization card.

1. The production authorization card signals that production of an inventory item can begin. This kanban usually lists the product's name, identification number, description, and the materials required in its production. It may also contain information on where materials or inventory can be found, and even component assembly information. In computer-based environments where work instructions for manual effort are provided at work centers by computer terminals, the kanbans can contain computer keywords to reference the instructions.

2. The vendor authorization card is used to signal a vendor to send the purchaser some specified units of supplies, materials, or inventory. This kanban usually lists the purchaser's inventory item name, the vendor's product's name, identification numbers, and an order quantity.

3. The conveyance authorization card is used to signal that a material handling person is authorized to move or withdraw supplies, materials, or inventory from a specific location to a specified destination. This kanban usually lists the product's name, identifying number, the location where the item should be picked up, and the location where the item is to be delivered.

A kanban system's operation is fairly simple. The issuance of one kanban card causes the production, vending, or conveyance of one unit of the desired product; the issuance of two cards causes production, vending, or conveyance of two units, etc. Ideally suited for a JIT environment, kanbans are issued on a daily basis permitting a rapid response to shifts in customer demand requirements. Not all three types of kanban cards need to be used to have a kanban system. Some organizations use a single card system, and others use a dual card system [8]. Regardless of which types of cards are used, they serve to authorize the production, purchase, and movement of inventory throughout the organization.

To successfully use kanbans an organization should meet the following requirements:

1. have a fairly stable demand of the finished products the system produces,
2. have a continuous flow production type of operation,
3. have a willingness to allow some WIP to exist in the system as a prerequisite to start up, and
4. have supplies, materials, and inventory items stored in single item, reusable containers (i.e., trays or boxes). This means that a tray will contain all of one type of component part used to manufacture a product.

While kanban systems can be used in situations that violate these requirements, the best results occur where these requirements are strictly observed. Kanban systems can be and are used in small and limited lot production environments [9].

In a kanban system, the cards are used to initiate transactions. The production, vending, and conveyance of items are the transactions. In a single card system only the conveyance card is used. Once a single card system is in place, it is easy to add a production or vendor kanbans to the system.

The Single Card System

To illustrate the single card system let's look at an assembly line work situation depicted in Figure 3-7(a). A worker in a work center needs inventory to complete a product. A conveyance kanban is issued from the work center defining the inventory required. The kanban is then placed in an empty container at point **a** in

Figure 3-7(a) This kanban acts to notify the material handlers that inventory is required and they are authorized to obtain it from the inventory storage department. A material handling person moves the empty container to the inventory department and drops it off at point **b**, keeping the kanban. The material handler then picks up the desired inventory from the full containers at point **c**. Note, there must be some WIP inventory or excess stock waiting for pickup, or the material handler will not be able to use that particular conveyance kanban. From point **c** the material handler moves the full container to point **d** in the work center where it can be processed by the worker. The material handler then goes back to point **a** to begin the cycle again.

FIGURE 3-7 • *Kanban Card Systems Inventory and Card Flow*

(a) Single Card Kanban System

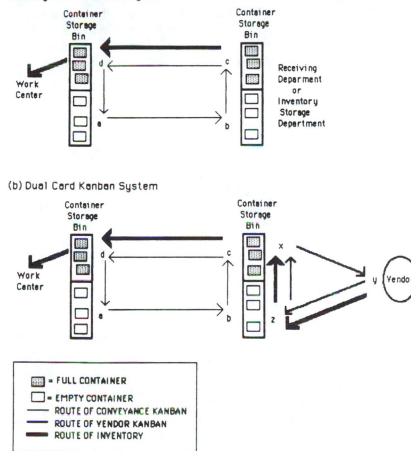

(b) Dual Card Kanban System

■ = FULL CONTAINER

□ = EMPTY CONTAINER

—— ROUTE OF CONVEYANCE KANBAN

—— ROUTE OF VENDOR KANBAN

▬▬ ROUTE OF INVENTORY

This single card system works well as long as there is some excess inventory available for pickup in the inventory bins. Since the kanban is based on a continuous flow production system, the start up or excess inventory can be reduced by eliminating some of the kanbans once the system is under way. In the single card example, if we had required three containers of inventory, we would have issued three kanbans. If we had an ongoing kanban system and wanted to eliminate some of the excess inventory in the system, we might issue only two kanbans when in fact we need three kanbans for the particular order. The effect on the operating system will be a reduction of one container of WIP inventory. We cannot issue kanbans unless excess WIP inventory exists in the system.

The single card system is most appropriate in repetitive operations where the same component parts are made by the same workers every day. The system works best with standardized, unitary, or limited lot JIT usage containers. They are standardized to the lot size of the product that is being produced. If we produce, for example, one radio at a time, then a container may contain only the parts to manufacture one radio. If on the other hand an organization feels that five radios are a more economical production lot size, then the container must contain parts for five radios.

The Dual Card System

In a dual card system we use two or more types of kanban cards. To illustrate the dual card system let's look at the assembly line work situation depicted in Figure 3-7(b). In this operation the work center obtains its materials from a vendor. The conveyance kanban is used in the same manner as in the single card system (i.e., the conveyance kanban travels from a, to b, to c, to d, and back to a). A vendor kanban is introduced at point x in Figure 3-7(b). The vendor card authorizes a vendor at point y to obtain and deliver specified inventory or materials defined on the vendor kanban card. Once the inventory or materials are delivered at point z, they may have to be broken down into a prescribed container unit by the vendor, who sorts them into several empty containers. Many vendors who service kanban organizations provide this materials sorting activity as part of their service. An empty container is obtained at point z by the vendor for materials storage. The filled container is then placed in the bins at point x where it will remain until a material handler will bring a conveyance card from point b to point c to authorize its movement. Once the container is authorized to be moved, its vendor kanban is removed and sent to the vendor where the system repeats itself.

We can see in Figure 3-7(b) that the two separate kanbans are in closed loops and continue to circulate until they are withdrawn by management. The issuance of the kanbans on a daily basis can support production control and inventory control when used in a JIT operation.

A kanban system has been very successfully used by many organizations [10]. One of its chief limitations, though, is that it is very dependent on the participants in the production operation. If participants lose cards or purposefully fail to perform their respective duties, management will lose control and the system may completely fall apart. Fortunately, in a kanban system, a card audit can be performed as quickly as each manager can count the respective department's cards.

Determining the Number of Kanban Cards

As previously stated, one of the requirements necessary for a kanban system to successfully operate is having some work-in-progress in the system. This WIP comes about by issuing some appropriate number of kanbans at each work center in the operation. An appropriate number of kanbans at one work center may not be appropriate at another work center because of the differing nature of work at different work centers, unit production usage rates required at different work centers, possible inefficiencies unique to a particular work center, and capacity limitations of containers (where multiple units are stored in the same container).

A formula that can be used to determine the ideal number of production authorization kanban cards to issue at each work center to support production usage is presented below [11, pp. 275-279]:

$$n_p = \frac{(d)(t)(1+e)}{c}$$

where n_p is the number of kanban cards to issue and be in use to support a specific production rate, d is the planned average daily unit production for the work center, t is the average unit (or lot) time for the setup or production expressed as a percentage of the day, e is a value that ranges from 0 to 1 that represents the percentage of inefficiency that exists in the system (i.e., a value of 0 means there is no inefficiency), and c is the unit capacity of a container (usually equal to 1, except in lot size production). In situations where the value of e cannot be calculated, or where there is a need for some inventory buffer stock, e would equal the desired level of buffer or safety stock expressed as a percentage of average daily demand.

A formula for determining the number of vendor authorization cards can be expressed as follows:

$$n_v = \frac{(d)(2+t)(1+s)}{(D)(c)}$$

where n_v is the number of vendor authorization kanban cards to issue and be in use to support a specific production rate, d is the planned average daily unit of production for a specific inventory item being vended, 2 is the minimum number of one-way scheduled trips required to complete a vendor transaction (i.e., one card pickup and one delivery), t is the transit time between picking up a card and making a delivery expressed as a percentage of the day, s is a buffer or safety stock level for the inventory item expressed as a percentage of daily demand, D is the number of deliveries per day currently scheduled, and c is the unit capacity of a container (usually equal to 1, except in lot size production).

A formula for determining the number of conveyance authorization cards can be expressed as follows:

$$n_c = \frac{(d)(t)(1+s)}{c}$$

where n_c is the number of conveyance kanban cards to issue and be in use to support a specific production rate, d is the planned average daily unit of production, t is the average material handling time expressed as a percentage of the day, s is a safety stock level for the inventory item expressed as a percentage of daily demand, and c is the unit capacity of a container (i.e., usually equal to 1, except in lot size production).

To illustrate the use of the production kanban formula, let's look at an example problem. Suppose we wanted to determine the number of production kanbans for a specific work center that will be expected to produce an average of 1,500 units of product per day, with an average unit (i.e., material handling, setup, and production) time of 5 percent of the day. Given an inefficiency index from time studies of 4 percent and a 10-unit capacity on the reusable containers, how many kanbans should be issued to the work station to support the daily demand rate of 1,500 units?

The answer is found by simply plugging the values of $d=1{,}500$, $t=0.05$, $e=0.04$, and $c=10$ into the kanban formula to obtain the desired number of kanbans:

$$n_p = \frac{(1.500)(.05)(1+.04)}{10}$$
$$= 7.8 \text{ kanbans}$$

To support an average daily production of 1,500 units in this work center, we should have 7.8 kanbans in use in the system at any one time. If we choose to round up to 8 kanbans in this example, we will be loosening the system with excess work-in-process. This may cause waste by generating unnecessary WIP. If we choose to round down to 7 kanbans, we will be tightening the WIP and stressing the amount of WIP available for use in other subsequent work centers. This may cause delays in material handling at subsequent work centers unless the system already has an excess of WIP inventory. There is no easy rule of thumb to guide users on which direction they should choose in this rounding decision. Management must balance, as always in JIT operations, the costs of kanban shortages with the costs of surpluses to decide on the exact number of kanbans to issue. The production formulas above only provide a starting point or initial guideline to begin the process of continual experimentation and improvement.

New Technology Supports Kanban Systems

Most U.S. manufacturers plan production schedules using computer-based systems. The significant investment and informational value of computer-based production planning is such that it tends to discourage the introduction of non-computer-based planning systems. The kanban systems have been and can be used without the need for any computer support. The issuance of kanban cards takes care of the scheduling activities so no computer support is viewed as essential for its successful operation. U.S. firms adopting JIT have been reluctant to embrace the supporting kanban systems because of their non-computer orientation. Being able to track orders on a timely basis is an important reason why U.S. manufacturers continue to use and stress the importance of computers in servicing the needs of their customers. Unfortunately, the time that would be wasted in recording kanban transactions in each production department's computer system runs counter to the JIT principle of avoiding wasteful inspection or recording time.

Some U.S. firms are finding new ways of integrating kanban systems with computer systems [12]. One new technology that allows for a timely recording of kanban transactions is with a bar coding system. **Bar codes** are electronically interpreted labels used to identify the contents of boxes or cans. In a kanban system, cards are used to define transactions that are being processed at work centers by material handlers and by vendors. These kanbans define very specific production, vending, and conveyance transactions that can be defined within a computer system. The transactions can be identified using a different bar code for each type of possible kanban issued. At the same locations in a production facility where the kanban transactions are completed (e.g., inventory storage bins, work centers, etc.), convenient optical scanners are positioned for use by all the participants in the kanban system (i.e., workers, material handlers, and vendors). By requiring participants to run their kanban cards quickly over a scanner upon the completion of their kanban transactions, the computer can interpret the bar code and record the transaction. This will permit the computer's information system to keep track of the production, inventory, material handling, and vendor transactions on a real-time computer basis. This type of integration allows the kanban system to operate with little effort to enter data into the computer system. This joint system also permits the computer to do what it does best, process large amounts of data quickly and provide timely information to managers.

The decreasing cost of bar coding technology and computer software to support informational needs of JIT operations is helping to convince U.S. manufacturers to adopt JIT methods like kanban systems [13]. We will be discussing bar coding technology in other chapters because of its many contributions to the implementation of JIT principles.

Measuring JIT Production Management Performance

Measurements of JIT production management efforts are necessary to demonstrate the improvement taking place with JIT production principles. There are several measures that can directly or indirectly measure production performance that can be used to evaluate JIT management progress in an operation.

In production we focus chiefly on measures related to human and equipment production resources. To indirectly measure the **value added** (a product's perceived utility by the customer) to products by production efforts of the workers in a JIT operation, two formulas can be used:

$$\text{Value added} = \frac{\text{(Dollar sales) - (Materials' purchase costs)}}{\text{Number of workers}}$$

$$\text{Value added} = \frac{\text{(Dollar sales) - (Materials' purchase costs)}}{\text{Total cost of workers}}$$

The ratio for the value added where the denominator is the number of workers provides a rough dollar profit per worker value. Under a JIT operation this ratio should increase over time, signaling the workers are helping to add value to the product such that it is impacting profits in a favorable way. The other ratio provides an exchange rate of profit per worker cost dollars. Again, the larger the ratio, the greater the value added per dollar cost of production. These ratios are used on a comparative basis over a period of time. As an operation evolves to more fully embrace JIT production management principles, these ratios should increase to reflect the greater value added by the workers to the products they produce under a JIT system.

A more direct measure of JIT worker effectiveness can be established for operations that use pre-set **standard time measures** for worker performance. If worker efficiency increases as we adopt JIT production management principles, it demonstrates that the JIT approach is effective at increasing productivity in workers. One formula that can be used to measure worker efficiency is as follows:

$$\text{Worker efficiency} = \frac{\text{Standard worker hours allowed}}{\text{Total JIT worker hours used}}$$

Standard worker hours are the number of hours allowed as standard or expected hours for some fixed number of units completed. The denominator of total JIT worker hours is the actual worker hours used in the completion of the fixed number of units. So the ratio divides the expected worker hour usage by actual worker hour usage. We would expect the ratio to be above one if JIT production principles are effective at increasing productivity.

A similar efficiency ratio can be used to measure the effectiveness of machine utilization under a JIT production approach.

$$\text{Machine efficiency} = \frac{\text{Total run hours used for scheduled production}}{\text{Standard hours allowed for scheduled production}}$$

Total run hours in the numerator are the actual total number of hours the machine was run to produce some fixed number of units. The denominator is the number of hours allowed as standard for the fixed number of units completed. We would expect the ratio to be below one and as close to zero as possible if JIT production principles are effective at increasing machine or equipment productivity. Only time for work that was scheduled for production is included in this ratio. Machines that are not needed in a JIT operation are not run. This is consistent with the JIT principle of minimizing waste. If machine work is not needed, the machine in a JIT operation is not run to save labor time and to avoid the unnecessary inventory the machine would produce.

Other more accounting-oriented measures of the improvements caused by a move toward a JIT production system include the total direct costs for setup, labor, machines, materials, and tooling [14]. While these are considered as classical manufacturing overhead measures of an operation's performance (when compared on a per unit production basis), some experts are suggesting that JIT cost accounting, and the efficiency measures that are calculated from the cost accounting, should be differently based by focusing specifically on **JIT overhead** [11, pp. 390-396]. "JIT overhead" cost accounting focuses on two basic elements:

1. the overhead costs such as labor, equipment, etc. of the value added area in a plant that is used to support direct manufacturing activities (i.e., theactual space in a facility used for manufacturing), and
2. the remaining overhead costs such as unused space, administration, etc., that act as a burden and must be supported by the value added activities.

The first cost accounting element above focuses on measuring the allocation of resources to the service side of an operation that produces products. The second element of JIT overhead focuses on the allocation of resources to the non-service, non-value-added side of activities that are carried by the service side activities. It is quite apparent that this type of costing is very much focused on measuring the JIT principles of increasing manufacturing efficiency and minimizing waste.

Implementation Strategy for JIT Production System

A number of strategies have appeared in the literature for implementing a JIT production system. One implementation strategy requires a ten phase project plan [15]. These ten phases include:

1. Preparation and Commitment Top management must learn what JIT requires in terms of resources and how they stand to benefit from JIT systems. Top management must then make a commitment to implement JIT throughout the organization. All employees and managers must then learn their roles in a JIT system, and top management must motivate them to be willing participants in the change that will be necessary to convert to a JIT production system.

2. Systems Investigation Data must be collected on the necessary production activities required to manufacture products. All production system input (i.e., labor, materials, and equipment) and output (i.e., finished products, subassemblies, etc.) must be defined. Simple charting of production activities can be used to help understand the material flow and production effort. These charts can then be used to identify waste and needless complexity that can be removed. Additional steps at this phase might include reducing the variety of products and their options to achieve uniform product designs, and increasing the use of interchangeable parts to reduce inventory and complexity. These actions help to streamline product design efforts, production planning, and control, and simplify setup time reduction activities.

3. Formation of Lines This phase involves the designing of flexible and mixed model lines capable of balancing the customer demand needs with the need for mix fluctuations. The workforce size and the mix of types of labor must be determined to define the line flow, sequence the operations along the lines, establish work centers, assign worker tasks and tools. The driving force for the formation of lines must be cycle time requirements to meet customer demand.

4. Formation of Cells To form a GT cell or any type of production cell we must start with the products that it will produce. This phase consists of establishing part families, and subsequent cells and fabrication lines for their production. Parts that do not fit into any part family should be redesigned or subcontracted. Once the product or product families are defined, the necessary production system requirements from Phase 2 can be arranged to achieve cycle time goals. Minimizing distance and movement within the cell is a major objective.

5. Layout Design This phase involves the proper location of subassembly, assembly, fabrication, and machining lines. The objective of layout design is to minimize inventory traveling distance and material handling costs. In some situations this involves balancing the flexibility provided by human resources versus the less flexible, more expensive, but faster automated material handling systems. The use of U- or C-shaped cells is often used because of their usefulness in minimizing the distances between machines, increasing flexibility, enhancing group effort, and providing for better communication among workers.

6. Lead Time Reduction In this phase we seek to reduce wasted lead time in machine operations, setup, waiting, and conveyance times. We might start by identifying processes with long lead times and breaking the process activities down to identify wasted activities that cause lead time. A team of workers observing a video tape of themselves performing their jobs during a production process is sometimes a useful way of obtaining suggestions on wasteful lead time effort related to processes, methods, or use of tools. At this phase, activities might include cutting lot sizes to see the lead time impact on work centers, using shorter conveyance lines to cut conveyance times, and reducing work center areas to facilitate communication and conveyance times.

7. Building System Stability The objective of this phase is to achieve a quality of consistency or stability in both products and system through preventive maintenance. By providing preventive maintenance that keeps the production system operating, the quality of production and the efficiency of the operation remains high. Employees are asked to perform routine and preventive maintenance activities to ensure consistency in both what they produce and when they produce it. This helps the system to maintain a stable level of production capacity by avoiding machine, line, or cell downtime.

8. Integrating a Pull System It is necessary to integrate all of a JIT production operation to obtain maximum JIT benefits. One of the best methods of integrating a JIT production system is with a kanban card system. The pull signals generated by kanban cards are highly visible and act to motivate the production function in a JIT operation to take place. To implement a kanban card system requires a determination of locations for communication areas and stock points. Where computer control is desired to monitor inventory or control production activities, kanbans can be sent electronically to work centers via computer terminals.

9. Supplier Integration This phase might involve training suppliers on how to use kanbans, or training them on many of the other JIT aspects of the purchaser-supplier relationship discussed in Chapter 2. The objective is to bring the suppliers into the planning process so they will know their future role and what is expected of them.

10. Continual Improvement Gradual change of an operation toward a JIT production system will take continual motivation and training for all the people involved in system implementation. To help maintain and promote a climate for JIT improvement the organization should establish programs for ongoing education and training of employees, to help instill positive work attitudes, cultivate a group or team effort philosophy, and develop performance measures for achieving JIT goals.

It is important to know that JIT goals are being achieved for both management and employees.

The ten phases of the JIT implementation strategy above need to be customized to meet each of the organization's unique production requirements. To be successful, though, in implementing JIT production operation managers must also try to avoid some of the factors that lead to a JIT program failure. Managers must try to avoid accepting poor quality, permitting a lack of discipline, allowing labor restrictions that might rob productivity, and they must seek to eliminate jobs that do not permit rewards for improved motivation.

Actual Case Study

The Ferro Manufacturing Corporation of Madison Heights, Michigan, decided to convert some of its production operations into a JIT production system [16]. Ferro Manufacturing Corporation has operated since 1915 with manufacturing plants located in several states and annual sales exceeding $100 million. They produce automobile industry component parts, like seat products, door mechanisms, and latching systems.

To implement the JIT production system they chose a single product on a single line in one of their plants as a pilot project. (Commonly when introducing a new production system, like JIT, it is logical to experiment with small pilot areas before converting the entire operation over to the new production method.) Four production objectives were outlined for the pilot project: (1) smooth production flow, (2) eliminate material handling labor, (3) reduce setup costs, and (4) reduce WIP.

Consistent with the JIT production implementation strategy, the employees were trained in JIT principles and asked to help improve their individual jobs by making suggestions to achieve the project's objectives. The training and changes in the design of the production line for the product in the project were completed in only ten weeks. The JIT design for the product reduced what had been eight separate assembly lines down to only six U-shaped lines with a single common aisle. By the end of the 15th month, the observed JIT production benefits consisted of: (1) productivity increased by 46 percent, (2) scrap reduced by 67 percent, (3) rework reduced by 93 percent, (4) assembly line floor space reduced by 15 percent, and (5) total cost of quality reduced by 47 percent. Operators in the plant became self monitoring on their scheduling goals, efficiency, and quality standards.

Summary

In this chapter we examined JIT production management principles and methods. Consistent with these principles, the management of a JIT operation should seek a uniform schedule of production that defines on a daily basis a load level product mix that will maximize JIT benefits while achieving monthly MPS goals. A JIT production operation should also have a facility layout that permits scheduling flexibility to achieve the desired synchronized pull system necessary for all JIT operations. The JIT layout should be focused on JIT principles in using automation where practical, cutting lot size, cutting setup costs, and improving communication and control. A JIT operation should also seek to improve flexibility in workers and in what workers do for an operation.

A number of tactics and methods useful to help managers implement JIT production principles were presented in this chapter, including: load leveling, mixed model scheduling, kanban card systems, focused factories, GT cells, cross training, production quotas, and visual management. To ensure that the tactics and methods work, this chapter also presented a number of JIT production management performance measures useful in the evaluation of production efforts in a JIT operation.

The outcome of successful JIT production management effort is not just a product, but a quality product. Stated as a key JIT principle in Chapter 1, a JIT operation should "seek product quality perfection." The next chapter is devoted to defining what JIT principles of product quality are all about.

Important Terminology

Aggregate production plan an aggregated (i.e., all units of all products) production plan is used to match aggregate demand with aggregate production resources of labor, inventory and production capacity.

Automated handling systems any motor driven inventory conveyance system that is in part automated is an automated handling system, including assembly lines, robotic systems, automated storage systems, and automated retrieval systems.

Bar code A label made up of a series of alternating bars and spaces that is encoded by computers, used to identify an inventory item by using electronic or optical scanning equipment.

Carrying costs costs that are associated with carrying inventory in stock, including insurance, taxes, storage facility costs, etc.

Changeover the tasks, time, and cost of setting up the equipment and inventory for a production run.

Continuous flow production a type of production which flows continuously rather than being proportioned into lots.

Cross training training personnel to perform a variety of different jobs.

Cycle time amount of time it takes to complete one unit of a product or the amount of time between successive completions of a product.

Economic manufacturing quantity (EMQ) a type of analysis or modeling approach based on calculus that seeks to determine the cost-minimized lot size for a production run. In the basic EMQ model the cost functions that are minimized are those of "carrying cost" and "setup cost." This procedure is based on the same type of methodology as the "EOQ" modeling approach discussed in Chapter 2.

Electronic sensors electronic devices used to monitor and sometimes adjust production equipment or products. Electronic sensors generate data that are conveyed electronically to computer systems. The computer systems then make comparisons between preset levels of equipment operation or product quality to determine if adjustments are required.

Flow shop a type of manufacturing facility layout and system that organizes machines, work centers, and workers in order of the sequence of work activities or stages of production that are required to complete products. An example would be an assembly line layout or a group technology cell layout.

Focused factory a facility that is dedicated to producing a limited number of different products with a limited number of production processes.

Group technology (GT) a philosophy that identifies the sameness of parts, equipment, or production processes to take advantage of common setups.

Group technology cell a type of layout design that permits rapid changes in the sequence of production activities. This layout is usually focused on the manufacture of a specific product family and structured in a U- or C-shaped flow shop layout.

JIT overhead a new type of cost accounting for overhead expenses. Based on two types of overhead costs (value-added and non-value-added), this method bases the allocation of costs on space and management resource uses.

Kanban a Japanese card system used to schedule production, vending, and conveyance activities.

Load leveling a production planning strategy that seeks to establish a smooth daily production rate while allowing unitary production levels of each product in a product mix to change from month to month.

Master production schedule (MPS) a capacity and inventory supply plan for meeting demand requirements on a daily or weekly basis. It is a specific and detailed product unit plan for production requirements and inventory usage.

Mixed model schedule method a scheduling method used to determine a production schedule where different models of a product are to be produced by the same production facility or GT cell. This scheduling method provides a production schedule that sequences the individual product family model production.

Product family a group of products that share similar production requirements and/or inventory components. A product that permits options in its various forms might be considered a product family.

Production plan a general guide that defines total production for an extended period of time, such as a year.

Production quota a short-term unit production goal for a worker or group of workers for a fixed period of time.

Production schedule a detailed plan that defines unit production for specific periods of time, such as a week or even a day.

Pull system a system describing the use of raw materials and component parts inventory in the production of units of finished goods whose demand is known. This system is not initiated until the demand for the finished product exits and hence pulls the inventory through the production process to complete the product.

Push system a system describing the use of raw materials and component parts inventory in the production of units of finished goods whose demand is forecasted or estimated. Inventory is pushed through the production system to meet the forecast demand regardless of the actual demand for the finished product.

Robot a mechanical device used to pick up, move, or position inventory during the production process.

Setup costs costs that are associated with setting up a production run, including setup time, labor, scrap material, machine lead time, etc.

Standard time measures a standard amount of worker labor time that should be required to complete a unit of work. These estimates are usually derived from time studies by industrial engineers.

Standard worker hours the estimated number of hours of labor required by workers to complete one or more units of product. These estimates are usually derived from time studies by industrial engineers.

Total sequence time total amount of time it takes to complete an entire sequence of lot size runs for a given product mix. The total of all **cycle times** required to complete a sequence of production that will include at minimum one unit of each product in a product mix.

Uniform daily production schedule a day-to-day schedule of production where there is little or no variation in production quantities between days.

Value added a product's perceived utility by the customer who will consume it.

Visibility management a management principle that seeks to facilitate the observation of deviation from goals by restructuring plant layouts and providing information to workers on their productivity performance.

Work center a specific production facility usually consisting of one or more workers, robots, or machines. An area where workers do their work, store their tools, and inventory is consumed.

Discussion Questions

1. What JIT production management principles focus on flexibility? Explain each and how they seek to improve production operations with flexibility. Explain where and why they might increase and/or decrease costs.

2. What is a uniform daily production schedule? How is it different from an MPS? How is load leveling related to this schedule?

3. What is a synchronized pull system? Could there be a synchronized push system, also? Explain.

4. Is a JIT production system pro or con automation? Explain.

5. What is a JIT-focused factory focused on? Explain.

6. Why is a product family used in a GT cell layout? Explain.

7. Why do cutting production lot size and cutting setup costs go together in a JIT operation? Explain.

8. What is visibility management and how can it help a JIT operation? Give an example.

9. How can the mixed model scheduling method be used to determine a JIT production schedule? Explain.

10. What is the difference between single card and dual card kanban systems?

11. What new technology can support a kanban system in a JIT operation?

12. What is value added in a JIT operation? Explain.

Problems

1. What would the reconfigured GT cell look like for product C in Figure 3-3? Draw the new GT cell configuration with the appropriate work centers and workers.

2. A company produces two products: A and B. The cycle time for product A is 4 minutes and the cycle time for product B is 2 minutes. What is the minimum number of units necessary in a sequence of production using the mixed model scheduling method? What is the total sequence time? How many units of each product will be produced in an 8-hour day?

3. What is the unit order sequence for the products in Problem 2? How many times will the sequence be repeated in an 8-hour day?

4. A company produces two products: *A* and *B*. The cycle time for product *A* is 3 minutes and the cycle time for product *B* is 7 minutes. What is the minimum number of units necessary in a sequence of production using the mixed model scheduling method? What is the total sequence time? How many units of each product will be produced in an 8-hour day?

5. What is the unit order sequence for the products in Problem 4? How many times will the sequence be repeated in an 8-hour day?

6. A company produces three products: *A*, *B*, and *C*. The cycle time for product *A* is 1 minute, the cycle time for product *B* is 2.5 minutes, and the cycle time for product *C* is 5 minutes. What is the minimum number of units necessary in a sequence of production using the mixed model scheduling method? What is the total sequence time? How many units of each product will be produced in an 8-hour day? What is one of the unit order sequences for the products in this problem? How many times will the sequence be repeated in an 8-hour day?

7. A JIT manager is trying to determine the number of production authorization kanban cards that should be issued to support a given level of production in a specific work center. How many production kanban cards should be issued if: (1) it takes one hour out of an 8-hour day, which is 12.5 percent or .125 of a day's time to produce one container's worth of work at the work center; (2) the planned average daily unit production to meet demand at the work center will be 2,000 units, (3) a safety stock of 5 percent of average daily production is allowed; and (4) the container's capacity is 100 units?

8. How many production kanban cards should be issued if: (1) it takes .005 of a day's time to produce one container's worth of work at the work center; (2) the planned average daily unit production at the work center will be 1,000 units; (3) the work center has an inefficiency rating of 2 percent of average daily production; and (4) the container's capacity is only 1 unit?

9. A JIT manager is trying to determine the number of vendor authorization kanban cards that should be issued to support a given level of production in a specific work center. How many vendor kanban cards should be issued if: (1) it takes 2 hours out of an 8-hour day, which is 25 percent or .25 of a day's time to produce one container's worth of work at the work center; (2) the planned average daily unit production to meet demand at the work center will be 1,000 units; (3) a safety stock of 10 percent of average daily production is allowed; (4) the transit time for a delivery is .125 of a day's time, and (5) the container's capacity is 50 units?

10. How many vendor kanban cards should be issued if: (1) it takes 20 minutes of time to produce one container's worth of work at the work center; (2) the planned average daily unit production to meet demand at the work center will

be 1,200 units; (3) a safety stock of 8 percent of average daily production is allowed; (4) the transit time for a delivery is .10 of a day's time; and (5) the container's capacity is 12 units?

11. A JIT manager is trying to determine the number of conveyance authorization kanban cards that should be issued to support a given level of production in a specific work center. How many conveyance kanban cards should be issued if: (1) it takes one half hour out of an 8-hour day, which is 6.25 percent or .0625 of a day's time, to produce one container's worth of work at the work center; (2) the planned average daily unit production to meet demand at the work center will be 1,400 units; (3) a safety stock of 12 percent of average daily production is allowed; (4) the material handling time is .005 of a day's time; and (5) the container's capacity is 5 units?

12. How many conveyance kanban cards should be issued if: (1) it takes 45 minutes of time to produce one container's worth of work at the work center; (2) the planned average daily unit production to meet demand at the work center will be 2,500 units; (3) a safety stock of 1 percent of average daily production is allowed; (4) the material handling time is .004 of a day's time; and (5) the container's capacity is 20 units?

13. A company's current dollar sales are $1.2 million with a purchase cost for materials of $0.9 million and 450 manufacturing workers. After one year of using JIT production management principles the company's current dollar sales are $3.5 million with a purchase cost for materials of $1.8 million and 600 manufacturing workers. Using the value-added JIT production management performance ratio, has the use of JIT paid off for this company? Support your conclusion with the necessary statistics.

14. A company's current dollar sales are $4.3 million with a purchase cost for materials of $2.7 million and a total cost of manufacturing workers of $1.2 million. After one year of using JIT production management principles the company's current dollar sales are $6.5 million with a purchase cost for materials of $4.8 million and a total cost of manufacturing workers of $2.0 million. Using the value added JIT production management performance ratio, has the use of JIT paid off for this company? Support your conclusion with the necessary statistics.

15. A company's standard hours for a month's production is 2,000 hours and workers under a large-lot operation system actually take only 1,950 hours a month. The company adopts a JIT system and the actual hours drop to only 1,850. Using the worker efficiency performance ratio, has the use of JIT paid off for this company? Support your conclusion with the necessary statistics.

16. A company has just switched from a large-lot operation to a JIT operation. One of the key departments in the production facility of this company uses a large metal shaping machine. The machine is scheduled for production 100 percent of the working week. Under the large-lot operation the company allows a standard time for the shaping machine of 30 minutes per piece of metal. In a 40-hour week the large-lot operation usually completes 80 pieces of metal. During the first three weeks during which the company went to a JIT operation, they completed 70, 85, and 96 pieces of metal, respectively. Using the machine efficiency performance ratio, has the use of JIT paid off for this company? Support your conclusion with the necessary statistics.

References

[1]   J. B. Dilworth, *Operations Management*, McGraw-Hill, New York, NY, 1992, chapters 9, 13, 14, 15, and 16.

[2]   R. G. Murdick, B. Render, and R. S. Russell, *Service Operations Management*, Allyn and Bacon, Boston, MA, 1990, chapters 7, 8, and 11.

[3]   W. J. Stevenson, *Production/Operations Management*, 3rd ed., Irwin, Homewood, IL, 1990, chapters 6, 7, 9, and 13.

[4]   R. J. Schonberger and E. M. Knod, *Operations Management*, 4th ed., Business Publications, Plano, TX, 1991, chapters 8, 9, 10, and 11.

[5]   H. Hirano, *JIT Factory Revolution: A Pictorial Guide to Factory Design of the Future*, Productivity Press, Cambridge, MA, 1989.

[6]   M. M. Helms, "Communication: The Key to JIT Success," *Production and Inventory Management Journal*, Vol. 31, No. 2, 1990, pp. 18–21.

[7]   R. A. Esparrago, "Kanban," *Production and Inventory Journal*, Vol. 29, No. 1, 1988, pp. 6–10.

[8]   R. J. Schonberger, "Applications of Single-Card and Dual-Card Kanban," *Interfaces*, Vol. 13, No. 4, 1983, pp. 56–67.

[9]   G. Plenert, "Three Differing Concepts of JIT," *Production and Inventory Journal*, Vol. 31, No. 2, 1990, pp. 1–2.

[10]   J. H. Im, "How Does Kanban Work in American Companies?," *Production and Inventory Journal*, Vol. 30, No. 4, 1989, pp. 22–24.

[11]   K. A. Wantuck, *Just-In-Time: For America: A Common Sense Production Strategy*, The Forum, Ltd., Milwaukee, WI, 1989.

[12]   K. Joshi, "Coordination in Modern and JIT Manufacturing: A Computer-based Approach," *Production and Inventory Journal*, Vol. 31, No. 2, 1990, pp. 53–57.

[13]   B. Whitman, "Applications for Bar Codes," *Production and Inventory Management Review with APICS News*, Vol. 10, No. 9, September 1990, pp. 44–46.

[14]   G. J. Miltenburg, "Changing MRP's Costing Procedures to Suit JIT," *Production and Inventory Management Journal*, Vol. 31, No. 2, 1990, pp. 77-83.

[15]   I. Nisanci and A. D. Nicolli, "Project Planning Network Is Integrated Plan for Implementing Just-In-Time," *Industrial Engineering*, October, 1987, pp. 50–52.

[16]   "Ferro Case Study," *Production and Inventory Management Review with APICS News*, Vol. 6, No. 11, 1987, p. 115.

CASE 3-1

Scheduling a Mix-Up

The Jameson Component Company (JCC) of Little Rock, Arkansas, produces five different electrical component parts for use in industrial robots. JCC's major customers are located throughout the U.S., Japan, and Europe. The company has for years produced products of such high quality that virtually no other competitor could impact their market share for the five component parts they manufactured and sold. This was particularly true during the 1980s when industrial robot sales had experienced a major growth period. In the recession-oriented 1990s, though, competitors' products were starting to take sales way from JCC. After considerable market research efforts, JCC found that the lost sales were a function of service and not product quality.

JCC's production facility used for the manufacture of the five components is based on intermediate-lot production. Lot sizes of 50 and 100 units of each of the components are produced and shipped to customers. The company set economical manufacturing quantity (EMQ) lot sizes of 50 and 100 for their products based on a prior consultant's overhead cost based recommendations. No special equipment limited their lot size decision.

The unique nature of JCC's dominance in the 1980s market allowed them to require rather large lead times for their production of customers' orders. JCC would take a customer's order on a daily basis and then backorder them until that customer's order quantity for components reached the 50 or 100 units lot size. JCC simply took as much time as was needed to add the individual daily orders together for their customers until they totaled the desired 50 or 100 unit EMQ levels. This permitted JCC to benefit from some of the demand pull experienced by JIT facilities. Unfortunately, as JCC was finding out, customers eventually gave their business to competitors who provided a more timely delivery service. Sales dropped to an all-time low and JCC management knew they had to do something.

JCC management decided that they would change their large-lot scheduling of production to a smaller-lot JIT operation. They also wanted to be able to schedule the production of all of their components each day. This, JCC management felt, would permit smaller and more frequent shipments to customers, regardless of the variety of components ordered. This would also permit a more rapid response to the smaller, daily orders JCC was experiencing from their customers.

To help accomplish the transition to JIT, they needed to set up a new mixed component schedule for production. In Exhibit 3-1 the currently used large-lot daily

EXHIBIT 3-1 • *Monthly Production Schedule for the Five Component Parts*

Component Model No.	*Scheduled Unit Production for Each Day of the Month*									
	1	*2*	*3*	*4*	*5*	*6*	*7*	*8*	*9*	*10*
WS134	50	50	50	50	50	50				
WA546							100	100	100	
WE788										50

Component Model No.	*Scheduled Unit Production for Each Day of the Month*									
	11	*12*	*13*	*14*	*15*	*16*	*17*	*18*	*19*	*20*
WE788	50	50	50							
WK686				100	100					
WM122							50	50	50	50

production schedule for a backordered month's worth of orders is presented. As we can see in Exhibit 3-1, the month's total production of the 300 units of Component Model No. WS134 is produced in six production runs of 50 each. Under such a schedule customers would have to wait unit the last four days of the month to receive any units of Component Part WM122. Based on current time estimates, the cycle time for Component Part WS134 is 8 minutes, the cycle time for Component Part WA546 is 4.8 minutes, the cycle time for Component Part WE788 is 9.6 minutes, the cycle time for Component Part WK686 is 4.8 minutes, and the cycle time for Component Part WM122 is 9.6 minutes.

Case Questions

1. What method discussed in this chapter can be used to determine the desired schedule? Why use this method and not some other?

2. JCC wants to produce all five components each day. How many units of each product will be produced each day in order to achieve the monthly production goals? How will this approach to production provide better service to JCC's customers? Explain.

3. What is the minimum lot size production quantity for each component part using the methodology selected in Question 1?

CASE 3-2

Can They Band Together with Kanban?

The Seiri Corporation of Kansas City, Kansas, is a small business operation producing a private label specialty item for a major retailer. The wholesale price for this specialty item was in excess of $1,000 in the late 1980s. The enduring nature of the demand for the single product they produce permits a fairly stable and constant scheduling approach to be used to plan production. The major retail customer Seiri has been doing business with for over 20 years has agreed to purchase all of the units of the product Seiri produces so long as the retailer's brand name is placed on the product, and Seiri agrees to an exclusive production contract limiting sales to only that retailer.

During the 1980s Seiri used a large-lot production scheduling method. Based on this method, Seiri produced to meet their own master production scheduled amount of 10,000 units per month. The plant is currently laid out in a typical economic manufacturing quantity (EMQ) departmental configuration (similar in nature to that layout presented Figure 3-2a). A single component inventory item is required for the production of the Seiri specialty item. The component inventory is currently acquired from a single vendor. The inventory is sent first through an assembly department and then through a finishing department for completion. Once a large lot of items is completed, they are sent to the retailer's warehouse to be broken down into smaller shipments that are then sent to the retailer's stores all over the country.

In early 1991, the retailer decided that a change in the way they had been doing business with Seiri was necessary. The retailer wanted to eliminate their warehouse to cut their operating costs, and have Seiri take on the responsibility of shipping smaller, more frequent orders to all of their retail stores. They also wanted Seiri to be able to ship their orders during the month on a more timely basis than the single monthly shipment of the past. The retailer wanted Seiri's order response time to be lowered to at least weekly, and hopefully, to just a few days once the order was placed. In return the retailer agreed to guarantee a total monthly minimum order from all of their stores of 20,000 units. The retailer wanted Seiri to respond to their new service request within the next month so they could begin to plan the closing of their warehouse.

Seiri's management was pleased to hear they would be having a 100 percent increase in monthly sales, but they knew the need for a change in order responsiveness would require a change in their EMQ-based facility. They needed a facility that would permit greater production flexibility on a daily basis to respond

EXHIBIT 3-2 • *Flow Shop Layout of Seiri Corporation's Manufacturing Facility*

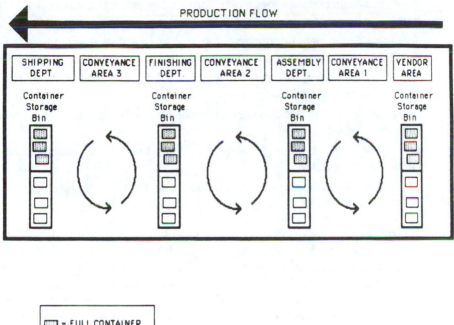

to individual retail store orders, and a facility that would still allow management some control on production flow. Being a small business, Seiri's management turned outward to a local production consultant for advice.

A production consultant from a local university was called in to advise them on what type of new production system they should adopt to best satisfy their customer's service needs. The consultant started by designing a new flow shop plant layout as presented in Exhibit 3-2. The component inventory in this layout is brought in at one end of the facility and moves through the facility in an almost straight line until it is shipped out the other end of the plant as a finished product. Unitary usage containers of inventory and WIP are temporarily stored in storage bins as they are moved through the plant by material handlers. The consultant also suggested that a dual card kanban system be used to band the departmental production, material handling, and vending efforts together while providing management with increased control in scheduling production in the new plant. Based on other similar plants that the consultant had designed, and known production lead time information from Seiri's existing production department activities, the consultant estimated the following:

1. Vendor information: the transit time for one round trip for the vendor is one hour, the vendor currently is planning 5 trips a day, and the container size is 1 unit.

2. Assembly Department information: the total setup and production time is 6 minutes per unit, and the container size is 1 unit.

3. Finishing Department information: the total setup and production time is 3 minutes per unit, and the container size is 1 unit.

4. Conveyance Area 1: average material handling time is 1.5 minutes, and the container size is 1 unit.

5. Conveyance Area 2: average material handling time is 3 minutes, and the container size is 1 unit.

6. Conveyance Area 3: average material handling time is 1 minute, and the container size is 1 unit.

To begin operations, Seiri's managemer. agreed with the consultant to allow a 20 percent buffer stock level as a planned safety stock for each of the six different areas in the plant. Based on this information, and the fact that the plant would have to produce an average of 1,000 units per day to meet the 20,000 unit monthly demand goal (assume 20 working days in the month), the consultant needs to plan the number of kanbans to issue in each of the six areas of the plant where data was collected.

Case Questions

1. How many kanbans should be issued to the vendor? Rounding the resulting value up, what assumption are you making?

2. How many kanbans should be issued to Conveyance Area 1? Rounding the resulting value up, what assumption are you making?

3. How many kanbans should be issued to the Assembly Department? Rounding the resulting value up, what assumption are you making?

4. How many kanbans should be issued to the Conveyance Area 2? Rounding the resulting value up, what assumption are you making?

5. How many kanbans should be issued to the Finishing Department? Rounding the resulting value up, what assumption are you making?

6. How many kanbans should be issued to the Conveyance Area 3? Rounding the resulting value up, what assumption are you making?

7. How could Seiri's management go about reducing WIP using the kanban system?

8. How does the kanban system give Seiri's management greater flexibility and control over production than when using an EMQ system?

CHAPTER FOUR

JIT Quality Management

Chapter Outline

Learning Objectives

After completing this chapter, you should be able to:

1. Explain what JIT quality management principles are and how they can improve a manufacturing operation.
2. Explain the importance in balancing job responsibility and job authority for workers in a JIT operation.
3. Explain how visibility management in JIT quality management is the same and is different from the visibility management in JIT production management.
4. Explain how a JIT operation can cost effectively achieve a 100 percent inspection.
5. Explain why continual improvement in quality over the long term is more important than quick profits in the short term.
6. Describe the difference between process control charts used to measure variables and those used to measure attributes.
7. Develop a series of process control charts for measuring and explain how they are used in JIT operations to improve product and process quality.
8. Explain what a quality control circle is and how it benefits a JIT operation.

Introduction

There is no single activity in business today that is more important to a company's survival than ensuring product or process quality. Since product quality is a relative concept, the question of how much quality is enough seems relevant. During the late 1970s and early 1980s, in the U.S. the common answer was to accept a small but allowable amount of poor quality in outgoing manufactured goods. It was felt at the time that the cost for reducing that small amount of poor quality was prohibitive. The Japanese during the same time chose a different course of action called "JIT quality management." Under this JIT approach product perfection was the goal and poor quality of any kind was not acceptable. The difference between the U.S. approach and the Japanese and other foreign manufacturers' approaches to product quality during that time can be evaluated to a large degree by the way in which foreign manufacturers have successfully captured many of the U.S. markets for manufactured goods. The pursuit of product perfection was the most important lesson that foreign manufacturers taught U.S. manufacturers in the 1980s, and it will continue to be a key element in successful business operations in the global markets where the U.S. must compete during the 1990s.

The manufacturing or service term, **quality**, usually means conformance to predefined product requirements. Quality in a product means different things to different people. Some of the more common requirements of quality are concerned with performance, features, reliability, conformance, durability, serviceability, aesthetics, and the concept of perceived quality. Quality control is the management function of verifying conformance to product requirements. In a JIT operation **total quality control** goes beyond simple control activities and embraces all phases of product quality. Total quality control involves all aspects of production effort, from product conception through work-in-progress to the delivery of the product. Its goal is to eliminate all contributors of nonconformity or defects in the production process [1,2,3,4].

The purpose of this chapter is to present the basics of total quality control as an introduction to JIT quality management principles. The principles that will be presented in this chapter are guidelines that when combined with the other previously presented JIT principles help cause the JIT benefits to occur from the **productivity cycling process** discussed in Chapter 1. As presented in Figure 4-1, product quality is a key element in both initiating the productivity cycling process and for the continual improvement necessary for success with a JIT operation. Successful firms that have adopted this JIT approach to quality did so by placing a higher priority on product quality than on a product's profit margin or unit production. This is not an easy step for some U.S. organizations that traditionally place the highest priority on profit maximization, but it is a necessary one if U.S. organizations hope to compete with Japanese and other foreign manufacturers that are generating superior quality products. In addition to JIT quality management principles, this chapter will discuss methodologies including process control charts and quality control circles. To help examine the success of a company in implementing JIT quality management principles, several measures of quality performance will also be discussed.

JIT Quality Management Principles

The basic JIT quality management principles include some of those necessary for total quality control [5, pp. 47-82]. The two goals of these principles are to achieve product perfection and to make quality control efforts a habit for employees. The JIT quality management principles include:

1. Maintain process control and make quality everybody's responsibility.
2. Seek a high level of visibility management on quality.
3. Maintain strict product quality control compliance.
4. Give the workers authority to share in the control of product quality.

FIGURE 4-1 • *Total Quality Control in the Productivity Cycling Process*

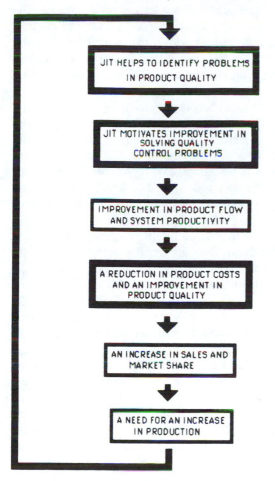

5. Require self-correction of worker-generated defects.
6. Maintain a 100 percent quality inspection of products.
7. Require workers to perform routine maintenance and housecleaning duties.
8. Seek continual quality improvement.
9. Seek a long-term commitment to quality control efforts.

1. Maintain Process Control and Make Quality Everybody's Responsibility **Process control** in manufacturing is the activity of controlling production process output within a given range of capability via some type of feedback or information system. Improving process control will result in improving product quality. Process control often involves the use of a collection of statistical techniques and charts to monitor and provide feedback for corrective control. The purpose of the **process control charts** is to provide graphic information on how well a production process is generating output within the range of designated capability. There are different charts to measure different aspects of product quality. Some of the most commonly used process control charts include: the x,- chart, the R chart, the p chart, and the c chart.

In Figure 4-2 a typical process control chart is presented. As we can see on a process control chart, the vertical axis shows a **quality standard,** which defines the desired quality level in the product in question, along with **upper control limits** (UCL) and **lower control limits** (LCL) which are boundaries that define expected deviation from the quality standard. The horizontal axis represents the time over which the production process's quality is observed. Plotted on the chart are samples of the production process's quality performance. The classic idea of process control charts is that the plotted sample quality performance should fall near or on the quality standards. The UCL and LCL act to provide limits on a process's expected deviation from a desired quality standard. They are usually being based on ±3 standard deviations from the quality standard value; the sampled output of process should form a **normal probability distribution** about the "quality standard" where 99.7 percent of the samples fall within the control limit boundaries. When these boundaries are violated, management should identify and solve whatever problem is causing the deviation. (We will be discussing the mechanics of several **process control charts** in a later section of this chapter.)

Before JIT came to the U.S., the UCL and LCL in process control charts were viewed as a static means of justifying acceptable deviation from stated quality standards. Deviation within the upper and lower control limits was statistically expected to follow the distribution of a normal probability distribution. A production process whose output quality fell between the control limit boundaries and was consistent with the spread of the normal probability distribution could be judged as producing acceptable quality. However, managers did not always wait until a boundary line was violated to investigate quality control problems. Most managers used the charts to discern patterns of deviation where the sample quality plots consistently fell in a direction heading away from the quality standard. Like a piece of equipment slowly going out of adjustment, the plotted samples of a process-generated product's quality would continue to deviate from a stated quality standard until management felt the deviation was unacceptably great, and corrective action was deemed necessary.

In today's JIT operations, the process control charts are used in a different and more dynamic manner. Process control charts are continually used to monitor deviation and call for corrective action from management. Any deviation from a quality standard is unacceptable and requires management's attention in a JIT environment where product perfection is the goal. Consistent with the desire to eliminate all deviation, the process control chart's upper and lower limits are used to express and motivate improved product quality. The wider the range between the control limits, the more variation in the production process, and the poorer the product quality. The control limits are viewed as dynamic boundaries that must be continually narrowed to achieve the JIT quality objective of perfection. In a JIT operation, the goal is to seek to reduce the width of the range of the control limits to zero. The goal of eliminating process variation is a very important concept in JIT quality management. Manufacturers continually compete on the basis of product quality. The customers who consume the manufacturer's products have their own type of quality control chart. A Japanese quality control expert, Genichi Taguchi,

FIGURE 4-2 • *Process Control Chart*

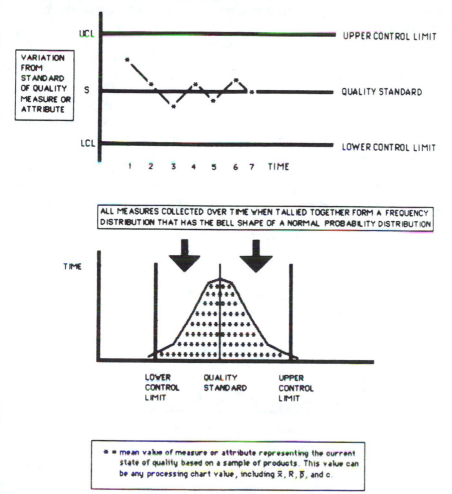

suggests that poor quality caused by process variation in production can generate a measurable loss for customers (and therefore manufacturers) [6]. Taguchi claims that a customer's loss due to a product's performance variation is often approximately proportional to the square of the deviation of the quality standard performance characteristic being measured. This type of loss function is depicted in Figure 4-3. Similar to the process control chart, customers have their own tolerance limits in which a product's quality is judged based on its nearness to a quality standard. The closer the product's quality is to the quality standard, the less the dollar loss incurred due to poor quality. Taguchi and JIT manufacturers in general know that the customer tolerance of process variation (i.e., the range between the customer's UCL and LCL) has been substantially reduced during the 1980s move toward improved product quality, and will continue to be reduced in the 1990s. To be successful a manufacturer must continually seek greater improvements in process quality.

Monitoring process quality with control charts will not cause product perfection by itself. Everybody who has a hand in making the product must have a hand in making it a quality product. In a JIT operation it is important to give responsibility to each person in the organization to perform quality control functions. For workers this might mean product inspection and quality control activities that might have previously been performed by quality control inspectors. It is quite common in a JIT operation that a worker at one work center is responsible for checking the quality of materials from a vendor. A worker might also be asked to perform product quality checks on another worker's value added efforts. This inspection of one worker's efforts by another places greater job responsibility on labor and acts to broaden the worker's job. This process is also

FIGURE 4-3 • *Taguchi Loss Function Due to Process Variation*

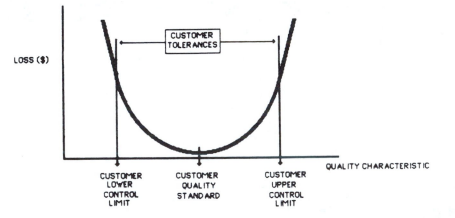

known as job enlargement and is commonly known to increase productivity and improve morale.

A worker's increased participation is not just limited to physical work tasks, but increased mental contributions should be encouraged as well. Workers should be motivated to offer ideas that might help to improve product and process quality. Workers with the responsibility to perform quality control activities are often motivated by the added responsibility to offer suggestions to improve product or process quality. This process should be more than just a suggestion box; managers must actively seek the counsel of worker participation in helping to solve production quality problems. One approach used by managers to enlist the help of workers in solving quality control problems is the use of quality control circles. (We will discuss quality control circles later in this chapter.)

In general, JIT management and workers must share the responsibility for quality control. All managers and their workers from the top of the organization to the bottom must be held responsible for their operation's production processes and their product's quality. To be successful, a JIT manufacturer must also continually improve process quality and the resulting products from those processes. By maintaining process control and making quality control everybody's responsibility, this principle marshalls an organization's resources to identify and correct product and process quality deviations. Embracing this principle focuses more human and informational resources on the critical issues of product and process quality control than in non-JIT operations. In turn, this JIT principle resource reallocation seeks to minimize dollar loss for poor quality while satisfying the 1990 customer's need for continually improved product quality.

2. Seek High Level of Visibility Management on Quality   In the past, management would keep process control charts to themselves. Why? Because management was held responsible for product quality. In a JIT operation, on the other hand, everybody in a plant is responsible for product quality. Everyone, therefore, should know the effects of his or her own unique contribution to the organization's product quality. Even more important for motivation purposes, everybody should know what everybody else's contribution is to product quality. Group pressure can sometimes be a greater motivator than any management dictate.

The process control charts are excellent graphic aids that describe worker, work center, or even plant quality behavior. In a JIT operation, workers are trained along with managers in the preparation and interpretation of process control charts. The use of control chart measures can for some firms have the same motivational effect as a batting average for a baseball player. If samples are taken daily for control charting, workers can see immediate increases or decreases in their efforts to change the quality of the product they are producing. Indeed, every behavioral scientist knows that the quicker the reinforcement of a behavior, the more likely the desired behavior will be repeated. Quality control charts can be placed on the walls or machines of work centers where everybody can see and observe quality control efforts. Similar to the visibility management of JIT production management discussed in Chapter 3, the visibility management of JIT quality management seeks to motivate change toward improving production quality.

One tactic that can help keep quality in the minds of production personnel is to assign the **quality control inspectors** as facilitators of JIT quality principles. Quality control inspectors can be used to educate and keep employees informed of the organization's progress toward quality control goals and new JIT innovations. Quality control staff personnel can also be used to train employees to understand how and why quality is monitored, and how to interpret the quality control charts.

3. Maintain Strict Product Quality Control Compliance In some sales-driven organizations more concern is given to meeting orders than to achieving higher quality. Quality control inspectors (who are traditionally staff members with no line authority) are sometimes pressured by managers (who are line members with firing and hiring authority over the quality control inspector) to let orders pass through the organization without necessary quality control inspections. Unfortunately, poor quality products rushed through an operation can end up costing more than the sales lost from not meeting a customer's order on time. Costs of poor quality can easily outweigh the profit on lost sales when they include: labor loss in the rework of defective items found during production, reprocessing labor loss in returned items from customers, scrap material costs when rework involves breaking down a defective unit, customer's perception of poor quality and resulting future lost sales, and rescheduling costs of rushing replacement orders for rejected products.

One tactic that can be used to help implement this principle is to permit a production rate of less than full capacity. It is important to identify and solve product quality processing problems. To ask the workers and management to do this will take some time away from productive activities and expected production capacity. By scheduling production at low full capacity, management allows time for the identification and solution of the quality control problems. The extra capacity allows for the production line to be shut down, keeps workers from overtaxing equipment (which can cause poor quality products), and may keep management from pressuring workers not to perform their quality control functions. The lessening of pressure on volume of production will also help demonstrate to all workers the greater commitment management has to quality production.

The JIT manager and workers must be motivated to think quality first and production rates second. A JIT operation should make quality a habit and insist all workers do their jobs right the first time. One tactic that might help to implement a quality-first attitude at a production facility could entail the establishment of a reward system (i.e., bonuses or pay increases) on product quality improvements. A tactic to ensure strict compliance with quality standards is to have **quality audits** performed every so often. By auditing the progress each work center, department, or plant is making toward stated quality standards, management is better able to find areas where increased JIT quality management efforts and resources should be allocated to effect a desired change.

4. Give the Workers Authority to Share in the Control of Product Quality

In some JIT operations workers and automated systems are given the authority to bring a production line, GT cell, department, or an entire plant to a halt if quality control problems are uncovered. Some plants use a more compromised system of worker signaling. This system consists of two lights: one light signals management that a minor problem has been discovered which could later cause a serious problem, and a second signals that the line should be stopped because a serious problem has been uncovered. In this type of system management can still retain the right to stop the production system. As a JIT rule, the amount of quality control authority that workers should be given should be dictated by the amount of quality control responsibility they are assigned by management.

This principle helps to reveal quality-related problems quickly, saving rework and scrap costs. While the cost of idle labor time for stopping a production line or GT cell might appear to be substantial under this principle, the workers are expected to fill their idle time by performing routine maintenance and housekeeping duties.

5. Require Self-Correction of Worker-Generated Defects

The worker who creates a defect in a product or the engineer who incorrectly designs a component part should be made to improve the specific quality characteristic whose standard they have failed to achieve. In this way, they will learn to do it right, and it may be an object lesson to other workers who will be motivated not to make mistakes because of the embarrassing consequences. Managers can also learn from this exercise. One of the benefits of this principle to management is that it can help to reveal problems that contribute to poor quality. Workers who are asked to correct product defects supposedly caused by their efforts often reveal a lack of skill or proper training to complete their production tasks. Managers, not workers, are at fault when workers lack the skills to do a job. Workers have also often been motivated to help identify quality problems related to poor materials, incorrect equipment use, and illogical production processes as a defense for the observed product defects they are being accused of causing.

6. Maintain a 100 Percent Quality Inspection of Products

Automobiles are made up of many types of materials, which are converted into components that are eventually assembled into a finished automobile. In a JIT automobile operation, each material and component part is inspected at each stage of production throughout the production process. The prior assembly of one component part at one work center is inspected at the next work center along a production line to make sure it meets pre-set quality standards. If one component part in a JIT operation is judged to be of unacceptable quality, it might keep an automobile from being completed. This process of inspecting the product quality at each stage of production continues to repeat itself until a finished automobile is generated. During the process, every piece of material, every component part, is inspected until the finished product is generally so defect free that a final inspection is not always even necessary. This critical principle actually builds quality into the product during the production process and has led Japanese automobile manufacturers to have some of the highest product quality ratings.

In actual practice, the quality control inspection work does not entail a great deal of duplication by workers along a production line. The first worker in a GT cell might perform a simple materials check to confirm vendor quality standards have been met. The second worker in the cell might just check the work performed by the first worker, the third worker checks the second worker's work, and so on. The only final checks that are often required in this type of operation are called product system checks to make sure entire systems of components are working together in accordance with quality standards.

This principle helps to quickly identify all types of quality problems. When coupled with the authority to stop production, workers motivate management to address and solve quality control problems quickly, thus minimizing their materials and scrap costs. Solving quality control problems helps improve production flow by eliminating future line stoppages that would be caused by the same problem. Solving quality control problems also saves labor time that would be consumed in rework and returns of poor quality production.

7. Require Workers to Perform Routine Maintenance and Housecleaning Duties In a JIT operation all workers should be assigned daily maintenance activities within their skill levels. Machine maintenance, such as simple oiling, tightening, inspecting tools to see that they are in shape, and replacing minor parts if needed are a part of the expanded job requirements of JIT workers. The five S's (i.e., seiri, seisetsu, seiso, seiton and shitsuke discussed in Chapter 1) are also a daily part of a worker's maintenance and housecleaning duties. Keeping the work center clean should be a daily habit for each employee. The maintenance and housecleaning work can usually be performed during production line down-times when quality problems are being corrected, so as not to cost the organization production time. Management will actually allow sufficient time during a day's operation to permit these duties to be performed by scheduling at less than 100 percent capacity as stated earlier.

This additional worker effort does not seek to eliminate either maintenance or janitorial personnel, only reallocate some of the more important quality production effort to the people who are now responsible for quality. The worker who uses a machine is in the ideal position to do minor maintenance work, since he or she knows best when machines need sharpening or oiling. On the other hand, maintenance personnel are best used to perform infrequent and more complex maintenance work like overhauls and year-end lubrications. The JIT worker is also in the ideal position to benefit from good housekeeping activities at the work center. Workers are more aware of the location of tools for efficient usage, and they will be motivated to keep tools orderly if they know they are responsible for putting them away at the end of the day.

8. Seek Continual Quality Improvement   The zero defect or goal of JIT is almost impossible to achieve, but is the ideal goal for all manufacturing operations. As long as process variation or product defects are found, quality control problems exist that need correcting. Finding the source of quality problems is a job of continual improvement.

One tactic used to help uncover problems in quality control is to cut lot sizes. Just as we did in implementing JIT inventory management principles, we can cut lot sizes to try and surface quality control problems quickly. Smaller lot sizes mean less inventory to hide problems. We want and invite product quality problems to surface as quickly as possible. Problems with vendor inventory quality, product engineering, and worker process training can all emerge as smaller lot sizes keep workers from hiding defects that larger lot sizes, with their built-in buffer stock, permit. Once problems surface, they can be corrected. If they remain hidden, for whatever period of time, the results are costly rework, increased scrap, and wasted labor time. This tactic can be strengthened by also cutting safety and buffer stock quantities within the production process. Without the extra stock, product quality problems are forced to surface.

9. Seek a Long-Term Commitment to Quality Control Efforts   To achieve a change in habit requires a long-term commitment. Moreover, it takes a proven track record of benefits to convince and motivate a continual commitment from everybody in a JIT operation.

As a company becomes a leader in product quality in an industry, its management will not find it difficult to remind workers of the company's success. For other companies planning to become industry leaders in product quality, one tactic to use is a project-by-project improvement record [5, p. 63]. By recording quality control project successes (and failures) that have been implemented at work centers, departments, or production facilities, a long-term record of what is possible in the area of quality improvement is established. The record should state what quality goals are sought, what is achieved, and what difference it makes for the company as a whole. This record should be made easily available and understandable to all workers and managers (consistent with the visibility management principle). It can be continually used by management to remind workers of the importance of product quality in their jobs and how it is making a difference in the long run.

The JIT quality management principles stated here are only introductory in nature. Each principle can be expanded to a study in and of itself. To help implement some of these principles we will examine a couple of the many quality management methods used to support JIT. One of these methods involves the use and development of process control charts.

Process Control Charts

Process control charts are used to ensure product quality during the manufacturing process. Information on product quality is obtained by inspecting a product's quality characteristics (e.g., height, length, defective features, etc.) for variation from stated quality standards. Some of these quality standard characteristics are measurable and some are not. Those characteristics that are measurable are called **variable quality characteristics.** Measurable characteristics include the weight, height, and length of a product (i.e., anything that can be measured and converted into a continuous numbering system). Other quality characteristics that cannot so easily be measured are called **attribute quality characteristics.** Attribute characteristics are concerned with product quality aspects such as function (i.e., an item that is defective or not defective) and appearance (i.e., acceptable or not acceptable).

In modern JIT systems many quality control inspections are performed by automated, mechanical, or electrical systems. Some of the automated process control systems can range in complexity from a scale used to weigh components to computer-based sensory systems using laser optical technology. Some of the process control inspection equipment is used to monitor the work-in-process (WIP), while other monitoring equipment is used to monitor machine performance. By monitoring either the product quality or equipment process quality the automated systems act to ensure the manufacturing process is kept under control. In Japan they refer to these mechanized quality control monitors as **foolproof devices,** since the human, who can be fooled, has been replaced with a machine that can almost never be fooled. Consistent with JIT quality management principles, some of these quality control machines can actually shut a production system down if the process quality drops below a pre-set level.

One method used to determine the control limits set in equipment to monitor manufacturing process deviation is **process control charting** (also called statistical **quality control [SQC] charts**). Process control charts are used to monitor quality based on **statistical confidence interval theory** and the **normal probability distribution.** These charts create a confidence interval about a quality standard or average level of quality a production process is generating. The confidence interval defines the boundaries of the upper control limit (UCL) and lower control limit (LCL) that are used to judge process control. The quality standard used in the charts is obtained by taking a very large sample of outgoing quality, or simply by using pre-set product specifications as a goal. For example, if we are filling 12-ounce cans of food, our desired quality standard amount of food to put into a can is 12 ounces. In this example the measurable variable representing the quality characteristic is the weight of food in a can. If we use a large sample, we could compute population representative variance statistics, like a **standard deviation,** to generate the control lines based on statistical confidence theory. The greater the number of standard deviations, the greater the amount of variation we are permitting. Most process control charts are based on ± 3 standard deviations from the quality standard. This represents an interval that should include 99.7 percent of all of the processing variation expected in the production processing system.

Why should any variation from the quality standard be allowed? In a JIT operation no variation of any kind goes unnoticed and uncorrected. Control charts in a JIT operation are used to monitor the variation in the manufacturing process and motivate workers and managers of the need for change to reduce the source of variation. Once a control chart's UCL and LCL lines are set up, additional samples are taken to see how the process variation is behaving over time. As we correct contributors (i.e., workers, machines, etc.) to poor quality, the reduction in product quality deviation from the quality standard will reduce the range in the control charts. On a periodic basis of a minute, hour, day, or week the control limits can be recomputed as needed by the organization. If everybody is using their JIT quality management principles, the quality control limits should continue to narrow over time. In this way the quality control charts support the visible management principle of JIT and improve communication to everyone concerning product quality.

To illustrate the use of this methodology, we will examine two control charts for monitoring variable quality characteristics and two control charts for monitoring attribute quality characteristics.

Control Charts for Variables

Two commonly used variable quality characteristics control charts are the bar chart and the R chart. The $\bar{x}$-chart is used to measure a production process's quality output variation from a quality standard that is related to the actual quality characteristic. In other words, the $\bar{x}$-chart is used to gauge deviation from a measurable variable, like the weight of the contents in a can or the length of a nail. The R chart is used to measure dispersion about a given level of desired dispersion. In other words, the R-chart's quality standard is a dispersion value that reflects variation. The $\bar{x}$ and R charts are usually used together since their joint interpretation of process quality complements their individual interpretations.

TABLE 4-1 • Confidence Factors for the A, B, and C Coefficients in the $\bar{x}$ and R Process Control Charts

Sample Size(n)	A	B	C
2	.880	3.267	0
3	1.023	2.575	0
4	0.729	2.282	0
5	0.577	2.115	0
6	0.483	2.004	0
7	0.419	1.924	0.076
8	0.373	1.864	0.136
9	0.337	1.816	0.184
10	0.308	1.777	0.223
11	0.285	1.774	0.256
12	0.266	1.716	0.284
13	0.249	1.692	0.308
14	0.235	1.671	0.329
15	0.223	1.652	0.348
20	0.180	1.586	0.459
25	0.153	1.541	0.459

The formulas used to calculate the UCL and the LCL for the $\bar{x}$-chart are presented in Figure 4-4(a). The formulas used to calculate the UCL and LCL for the R chart are presented in Figure 4-4(b). The values of A, B, and C in the formulas are found in Table 4-1. These values provide a ±3 standard deviation conversion of the average range to permit the 99.7 percent confidence interval. The simplicity of the table values also makes possible their use by workers who do not understand the statistical theory on which they are based.

The procedure for developing and using the $\bar{x}$ and R charts is as follows:

1. Determine the inspection sample size and the number of samples to collect The $\bar{x}$-chart is a chart where the measurable quality characteristic standard can be determined by a grand mean value. The **grand mean** is found by taking the average of the sample averages. We use the sample average values to plot quality on the $\bar{x}$ charts. So we must determine the number of averages, m, we will use to develop the chart and the sample size, n, for each sample. The R chart is a chart of ranges where the individual ith ranges or R_j's are plotted to reflect the variation that exists in the production process. The determination of n and m used in the $\bar{x}$ and R statistics can be based on factors such as the time or cost required to collect and perform the inspection efforts. The costs of inspecting units (management time, labor time, destructive test costs, etc.) must be balanced with the cost of inspection error (product liability costs, inaccurate information, costs of poor quality, etc.) in the final judgment on sample size. It must be remembered that the value of n is constant for all samples.

2. Collect the data for the process control chart's derivation This usually involves the selection of quality characteristics to measure, and the determination of the type of physical inspection of products needed to collect the data on the variable quality characteristics.

3. Compute the UCL and LCL values Using the formulas in Equations 4-1 through 4-4 we simply plug the appropriate values in and compute the desired UCL and LCL values for each chart. We assume that a 99.73 confidence level is adequate for monitoring purposes.

4. Draw the $\bar{x}$ and R charts The charts should clearly denote the S, UCL, and LCL lines. As a result of this step, the process control charts are ready for use until such time as they need to be revised. The timing of such revisions can be based on real-time individual completions of the production on an hourly, daily, or weekly basis. Since the use of these charts, though, is to monitor future production process output and the improvements in process quality, we must continue to Step 5.

Equation 4-1.

$$\boxed{\text{X-Bar Chart}} \quad \text{Upper control line } (UCL_{\bar{x}}) = \bar{\bar{X}} + A\,\bar{R}$$

$$\text{Lower control line } (LCL_{\bar{x}}) = \bar{\bar{X}} - A\,\bar{R}$$

Equation 4-2.

$$\boxed{\text{R Chart}} \quad \text{Upper control line } (UCl_R) = B\,\bar{R}$$

$$\text{Lower control line } (LCl_R) = C\,\bar{R}$$

Equation 4-3.

| P Chart |

$$UCL_p = \bar{p} + 3\sqrt{\frac{\bar{p}(1-\bar{p})}{n}}$$

$$LCL_p = \bar{p} - 3\sqrt{\frac{\bar{p}(1-\bar{p})}{n}}$$

Equation 4-4.

| C Chart |

$$UCL_c = \bar{c} + 3\sqrt{\bar{c}}$$

$$LCL_c = \bar{c} - 3\sqrt{\bar{c}}$$

WHERE:

$\bar{\bar{x}}$ = grand mean of all samples = $\dfrac{\sum \bar{x}_i}{m}$

m = number of samples

$\bar{x}_i$ = sample mean of the i th sample = $\dfrac{\sum x_i}{n}$

n = sample size

R_i = range of the i th sample

$\bar{R}$ = average range for all m samples = $\dfrac{\sum R_i}{m}$

A, B, C = tabled confidence values

$\bar{p}$ = mean fraction defective = $\dfrac{\text{Total defectives found}}{(m)(n)}$

c_i = number of defects found in each i th unit

$\bar{c}$ = average defects found per unit = $\dfrac{\sum c_i}{m}$

5. Take new samples of the production process at predetermined times, plot them and interpret their meaning The $\bar{x}$-chart is a chart of sample averages of output quality of a production process. New sample data is collected on the quality of the production process and converted into averages which are then plotted over time on the chart. The R-chart is a chart measuring variation. New sample ranges are computed and plotted over time on this chart. We continue collecting and plotting data until we feel the need to develop new charts. In a JIT operation the charts are also posted for review by workers and managers. Any observed deviations from quality standards are interpreted to focus attention and identify causal factors for poor quality from the particular work center, department, or facility the chart is measuring.

To illustrate the $\bar{x}$- and R-chart procedure, let's look at an example problem. Suppose we have a machine filling 12-ounce cans with soup. We want to develop $\bar{x}$- and R-chart to monitor how well the pouring of soup is going in the production process. In Table 4-2 the sample data for the problem is presented. In this problem we are taking three samples ($m = 3$), each consisting of an inspection where we measure the weight of soup in four cans ($n = 4$). The necessary statistics are computed here:

$$\text{Grand Mean} = \bar{x} = \frac{36}{3} = 12 \text{ ounces}$$

$$\text{Average Range} = R = \frac{4.4}{3} = 1.467 \text{ ounces}$$

$\bar{x}$-chart computations
Upper control line $(\text{UCL}_{\bar{x}}) = \bar{x} + A\,R = 12 + (0.729)(1.467) = 13.069$

Lower control line $(\text{LCL}_{\bar{x}}) = \bar{x} - A\,R = 12 - (0.729)(1.467) = 10.931$

R chart computations
Upper control line $(\text{UCL}_R) = B\,R = (2.281)\,(1.467) = 3.347$

Lower control line $(\text{LCL}_R) = C\,R = (0)\,(1.467) = 0$

As we can see, the quality standard $S = 12$ ounces is, as we expected, the production process average (i.e., the grand mean). (If the quality standard we desired does not equal the grand mean, we can choose to arbitrarily set the grand mean at the required 12-ounce level or wait to implement the charting systems until we correct quality factors in the production process that are causing the deviation away from the 12-ounce quality standard.) The UCLs and LCLs are calculated by using the A, B, and C confidence factors from Table 4-1. Note that the LCL for the R-chart turns out to be zero. Since it is realistically impossible to have a negative LCL in an R-chart, the default value for negative LCLs will always be zero.

TABLE 4-2 • x̄- and R-Chart Example Data Computations

Example Data of Three Sample Days of $n = 4$

Sample Day No.	Ounces in Each Observation (x)	Σx	x_j	R_j
1	12.4, 12.0, 11.6, 12.4	48.4	12.1	0.8
2	13.6, 11.0, 12.6, 10.8	48.0	12.0	2.8
3	10.6, 11.6, 12.4, 13.0	47.6	11.9	0.8
			$\Sigma = 36.0$	4.4

We can plot the sample averages and ranges between the boundaries of their respective control limits to see what is happening in this production system. The x̄- and R-charts for this problem are presented in Figure 4-4. In most cases, as presented in Figure 4-4a, the x̄-chart samples fall as expected within the control limit boundaries. In interpreting the x̄-chart, it appears that the general direction of the sample averages indicates a downward deviation over time toward the LCL. The chart's interpretation leads us to conclude that a tendency to package a reduced amount of soup in each can should be investigated.

The difference in content between the x̄- and R-charts for this example explains why these two charts are often combined to measure process variation. The x̄-chart's variation behavior is not as well captured as in the R-chart. Note the unusual variation in samples that can be observed in the R-chart in Figure 4-4(b). Despite the minor deviation from the standard in the x̄-chart in Figure 4-4(a), the R-chart's measures appear to be violently changing from sample to sample. While the three samples do fall within the boundaries of the R-chart control lines, their violent upward and downward movement indicates much variation in the production process. Ideally, the R-chart standard should be as close to zero as possible. The fact that so much variation exists in this R-chart should necessitate the identification of the production process factors that contribute to the observed variation. These sources of variation, once identified, should be reduced or eliminated to improve product quality.

The data used to generate the upper and lower control lines can be considered historic data. Generally, once the control charts are prepared, only new sample data is used to identify deviations from quality standards. In a JIT operation, nothing should be wasted including historic data. JIT managers should identify and correct any aspect of their production operation that contributes to poor quality based on past, present, or future data.

FIGURE 4-4 • x̄- and R-Charts for Example Problem

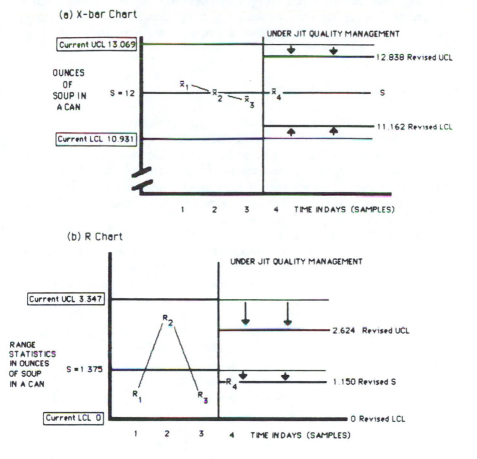

(a) X-bar Chart

(b) R Chart

Let's continue with this example and assume that the firm adopts JIT quality management principles. Let's further assume that the firm starts solving quality related problems in their operation at the end of Day 3. On Day 4, a new sample is taken of four cans of soup resulting in weights of 12.0, 12.1, 11.9, and 12.0 ounces. In a JIT operation the impact of quality improvements should be communicated as soon as possible to reinforce quality as a habit. If we add the additional sample into the computations of the existing UCL and LCL values for both charts, their boundaries will be revised as presented in Figures 4-4(a) and (b) for Day 4. As we can see, the UCL and LCL have been narrowed, reflecting an improvement in quality by a reduction in process variation. Workers at work centers can easily see and understand that the narrowing of boundaries is a direct reflection of their efforts to improve quality, that their efforts are making a difference, and that there is still more variation needing reduction. Using the process control charts as motivators to communicate the need for reducing variation, rather than as an excuse to accept variation as in the past, is the major difference between the Japanese use of process control charts and their past classic use.

Control Charts for Attributes

Occasionally, we do not have quality characteristics that can be measured. Whether it is testing for defective light bulbs or flaws in product finishes, such attributes cannot be weighed or measured by degree. Fortunately, there are many different control charts that can be used to establish control limits for attribute quality characteristics. Two types of process control charts for attribute quality are the p-chart and the c-chart. The same five-step procedure presented for the $\bar{x}$-and R-charts can be used for the p- and c-charts.

A **p-chart** charts the fraction or proportion of a sample that is defective. This type of chart can be used to monitor process control in situations where the quality characteristic inspections result in each item being judged defective or not defective in a lot, or for a situation where a fixed number of checks are made on a single item. Hence, a fraction, or a p proportion of the items in a lot or checks on a single item, can be classified as defective. To use a p-chart we must: (1) have a finite population of units or products (i.e., a lot or batch of units) on which to compute the desired fraction of defectives, or a finite number of inspection checks on a single item of which some can be judged defective, and (2) have a two-outcome judgment of whether the product quality is defective or not defective.

If we sample several lots and average these p proportions or fractions of defectives, we derive an average proportion defective that can be used as a quality standard for the chart. This fraction defective standard is an average so we can also calculate the standard deviation from the average for computing control limits in much the same way as we did for the $\bar{x}$-chart. The formulas for the UCL and LCL for the p-chart are presented in Equations 4-5 and 4-6.

Equation 4-5.

$$\boxed{\text{P Chart}} \quad UCL_p = \bar{p} + 3\sqrt{\frac{\bar{p}(1-\bar{p})}{n}}$$

Equation 4-6.

$$P \text{ Chart} \quad \text{LCL}_p = \bar{p} - 3 \sqrt{\frac{\bar{p}\left(1-\bar{p}\right)}{n}}$$

Where:

$$\bar{p} = \text{mean fraction defective} = \frac{\text{Total defectives found}}{\left(m\right)\left(n\right)}$$

m = number of samples

n = sample size

The objective in a JIT operation is to have as small a fraction defective as possible. Like the R-chart, the default value for a negative LCL limit in the p-chart is zero. As with the other quality control charts, the JIT use of the p-chart is to motivate workers to reduce percent defectives as close to zero as possible by visibly depicting the need to reduce process quality deviation.

To illustrate the p-chart, let's look at an example problem. Suppose we want to establish a p-chart for a television manufacturing company. The company has decided that it can only afford to make 100 special inspection tests of components (i.e., sample size n=100) on a single television set each hour for eight hours (i.e., number of samples m=8) in a single day for the purpose of establishing the control chart limits. The tests will be evaluated on a two-outcome basis of components being either "defective" or "not defective." The inspector will record the number of defective components found out of the 100 special inspection tests for each television set as a means of determining the average number of defects per set the television production process is generating. The collected data for this example is presented in Figure 4-5(a) and the control chart computations are presented in Figure 4-5(b). The resulting eight samples of 100 tests each generated a mean fraction of defective components of 0.04, which becomes the quality standard, S, for this television manufacturing process. The resulting UCL for the fraction defective is 0.098, and the LCL is defaulted to a value of 0 (i.e., the actual LCL is an impossible value of -0.18).

When plotting the eight proportions of defects used to make up the chart, the behavior of the production process becomes obvious. The proportion of defects is decreasing in a very specific and desirable direction. JIT managers should use this information to find out what is happening in the production process that is reducing defects in the later part of the day (i.e., the 7th and 8th hours of the work day). Learning what is done right to reduce product quality variation in a production process is just as important as learning what is done wrong. JIT managers can also use control charts to motivate workers by posting them in places where workers can see their quality control efforts are making a difference. The **c-chart** is similar to the *p*-chart in that it is used to chart attributes. The *c* in a *c*-chart stands for *counting*, because this chart is used when we want to establish control limits based on counted defects. With the *c*-chart, though, we do not have to have a finite

FIGURE 4-5 • *P-chart Example Data, Computations, and Chart*

(a) Example Data of n = 100 Inspections Per TV Set at One Set Per Hour for m = 8 Hours

Sample Hour No.	Defects Out of 100 Inspections Per Set	Proportion of Defects
1	7	.07
2	7	.07
3	6	.06
4	5	.05
5	3	.03
6	2	.02
7	1	.01
8	1	.01
	$\Sigma = 32$	

(b) P Chart Computations

Mean fraction defect = $\bar{p} = \dfrac{32}{(8)(100)} = 0.04$

P Chart

$$\text{UCL}_p = \bar{p} + 3\sqrt{\frac{\bar{p}(1-\bar{p})}{n}} = 0.04 + 3\sqrt{\frac{0.04(1-0.04)}{100}} = 0.098$$

$$\text{LCL}_p = \bar{p} - 3\sqrt{\frac{\bar{p}(1-\bar{p})}{n}} = 0.04 - 3\sqrt{\frac{0.04(1-0.04)}{100}} = 0$$

(c) P Chart

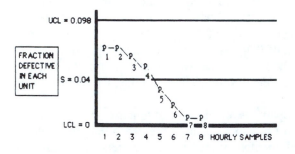

number of possible defects in the item or lot under study. For example, if we are inspecting a paint job, we might find zero flaws or some infinite number of flaws. The c-chart only requires that the defects are countable. In a c-chart the value of the average number of counted defects becomes the quality standard, S, for this control chart. The formulas to compute the UCL and LCL for the c-chart are shown in Equations 4-7 and 4-8.

Equation 4-7.

$$C \text{ Chart } UCL_c = \bar{c} + 3\sqrt{\bar{c}}$$

Equation 4-8.

$$C \text{ Chart } LCL_c = \bar{c} - 3\sqrt{\bar{c}}$$

Where:

c_i = number of defects found in each ith unit

$\bar{c}$ = average defects found per unit = $\dfrac{\sum c_i}{m}$

Like the R-chart, the default value for a negative LCL limit in the c-chart is zero. As with the other quality control charts, the use of the c-chart is to motivate workers to reduce the number of countable defects as close to zero as possible by visibly depicting process quality deviation.

To illustrate the use of this control chart, let's look at an example problem. Suppose we inspect paint jobs on automobile components and want to set up a control chart to monitor the quality of the paint job each component receives. The control chart is to be set up to monitor the observed number of flaws in the paint job of each component. Obviously, we cannot use either an $\bar{x}$- or an R-chart because we are not measuring flaws, only counting them. Likewise, we cannot use a p-chart because there is no way to know how many possible flaws can exist on a component, so a fraction defective cannot be determined. We can, though, use a c-chart because flaws are countable.

Suppose we want to establish the chart's control limits by taking a single component each day for five days, and then counting the number of paint job flaws observed in the inspection. The data for this example is presented in Figure 4-6(a) and the c-chart computations are presented in Figure 4-6(b). The quality standard represented by the average number of flaws is 4, with a UCL of 10 flaws and LCL default of 0 flaws. From the resulting c-chart in Figure 4-6(c) it appears the number of flaws is increasing over time, and management's corrective action is clearly needed before the UCL is violated.

All four of the control charts presented in this section suffer from inherent inaccuracies that exist in any statistical procedure, including inaccurate data collection and sampling error. Indeed, an underlying assumption made for all of the control charts is that their distribution is normally distributed. In most cases where a large enough number of samples are taken over a long period of time, this assumption usually holds. In actual practice we would have taken a larger number of samples with a larger sample size to establish the initial control limits. In fact,

FIGURE 4-6 • *C-chart Example Data, Computations and Chart*

(a) Example Data of the m = 5 Paint Job Inspections, One Per Day

Sample Paint Job No.	Defects Observed Per Paint Job
1	1
2	2
3	4
4	5
5	8
	$\Sigma = 20$

(b) C Chart Computations

Average defects per paint job = $\bar{c} = \dfrac{20}{5} = 4$

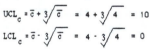

$$\boxed{\text{C Chart}}$$

$$UCL_c = \bar{c} + 3\sqrt{\bar{c}} = 4 + 3\sqrt{4} = 10$$

$$LCL_c = \bar{c} - 3\sqrt{\bar{c}} = 4 - 3\sqrt{4} = 0$$

(c) C Chart

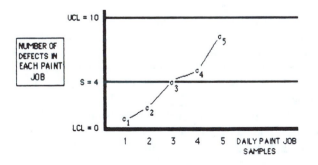

the charts are often used in JIT operations on a dynamic basis where each unit's completion has its quality characteristics recorded and used to adjust the control limits on a real-time basis. Old data is dropped out as new data under a JIT system is added. This permits workers to more rapidly see the impact of their quality efforts.

Inspection procedures, like process control, help to identify poor quality in components or products after they happen in the manufacturing process. It is well known that we cannot put quality into a product by screening out defective units. Using process control charts helps to motivate JIT workers to seek out and reduce production process variation that causes poor quality products. To help solve the problems found in a JIT operation, a group decision-making effort is needed by workers, managers, and technicians. One such group decision-making method is called quality control circles.

Quality Control Circles

A **quality control circle (QCC)** is a small group of workers, managers, and technicians assembled to study quality control problems and suggest solutions. A QCC is used to solve all types of quality control and production problems. The QCC is usually a formal and small group of interested personnel who are brought together as a team to study problems that each has some vested interest in solving.

The basic idea for QCCs came from small, informal groups of workers meeting to discuss mutual problems during coffee breaks and after working hours. The formal concept of QCC originated with Toyota, Inc. of Japan in the early 1960s when small groups of workers were assembled to help train other workers on quality control procedures and to deal with small scale quality control problems. Toyota refers to QCCs as **small group improvement activities (SGIAs)**. Not surprisingly, worker morale of these group members is greatly improved because of their ability to participate in a group decision-making process. Today QCCs are used to solve many different types of problems. Their objectives include more than just improving product quality, but also improving work methods, worker morale, and motivation [7].

QCCs differ in composition from organization to organization. In general they are comprised of workers, management personnel, and technical specialists like **value engineers, industrial engineers,** and quality control inspectors. The technical specialists are useful in their roles of improving work methods, redesigning tools, developing new work center layouts, and conducting work studies to improve worker efficiency. QCCs are not usually given the authority to implement their problem solutions; they just develop a set of recommendations for a final decision by the manager who forms the circle.

A typical QCC program development and implementation might involve the following steps:

1. A manager in the organization identifies that a quality control problem or production problem exists.

2. The manager establishes a formal agenda that defines the boundaries of the problem to be studied, who should be members of the QCC, and the timing of the meeting of the QCC on a daily, weekly, or monthly basis. The manager assembles the members from several types of organization employees, including: (1) a few workers who are assigned work in the problem area, (2) one or two supervisors who have supervisory responsibilities in the problem area, and (3) one or more technical specialists selected on the basis of perceived background and how they can be used to solve the problem under study (e.g., if we are studying a worker motivation problem, we might want to have an industrial psychologist as a team member). The managers in the QCC are not necessarily made leaders of the QCC, but are treated only as equal members when actively participating in the group decision-making efforts. The technical specialists can come from any staff function in the entire organization.

3. A QCC team leader is assigned to help coordinate the meeting times and keep the group targeted on the formal agenda.

4. At each meeting members of the group are free to comment and offer ideas on solutions to the problem. Both the workers and supervisors are equal members, and in the group situation are free to offer any relevant suggestion that might help deal with the problem under study. That is, the usual supervisor and subordinate relationship is temporarily subordinated to permit a freer environment to offer suggestions. Eventually, the problem is clearly defined, and a series of possible alternative solutions are developed. In some groups consensus methods are used to determine the solution strategy.

5. The problem alternatives developed by the QCC are usually offered to the manager who formed the circle for the manager's approval and final solution choice. Once the solution strategy is chosen, the same QCC may be asked to develop an implementation strategy.

6. When the problem is solved the QCC can be disbanded until another problem requires formation.

The benefits of this group problem-solving approach help not only the organization, but the individuals that make up the QCCs. The workers benefit by having a chance to express problems they experience to management and technical specialists who must listen to them as a part of the group. This helps achieve the JIT production management principle of improved communication. The team sessions also provide a break in job routine for the workers, and in some organizations the workers are given rewards (e.g., vacations, appliances, bonuses, etc.) for solving important quality control problems. The QCC also acts as a means to allow workers to achieve the JIT quality management principle of making quality everyone's responsibility. The technical specialists benefit by obtaining information on production processes and product quality from workers who have to implement the specialists' engineering theories in the practical world of the shop floor. Often this type of first-hand information does not make it through an organization's bureaucracy, since quality control problems are often different when communicated from the workers through a supervisor's point of view. Management benefits in many ways from QCCs. They will generally have more highly motivated workers because of the workers' greater job participation (the added QCC activity is a **job-enlargement** process). The workers are also more willing to accept change in work methods and physical plant layout when they have participated in planning them. The ideas that are generated from the QCCs also help to improve product quality, which in turn can reduce scrap and save production costs. Finally, management can use the QCC as an auxiliary problem-solving unit when they do not have time to deal with a problem themselves.

Measuring JIT Quality Management Performance

There are quite a number of ways to see that JIT quality management is making a difference in an organization's operation. Since we set as a goal that a JIT operation should generate a perfect product every time, we generally keep track of product imperfections, sometimes called nonconformities, as a means of measuring quality performance. A **nonconformity** exists when the output of a system does not conform to stated quality goals. Two ways to keep track of nonconformities are to count them or assess their costs.

Counting Methods

Defects found in WIP internally during the production process can be counted on a daily, weekly, or monthly basis as a direct measure of process nonconformities. Even the assessed range in the process control charts can be measured and counted as an internal indicator of process variation. Defects reported externally by customers can also be counted on a weekly, monthly, or yearly basis as a direct measure of finished product nonconformities. We would expect quality to improve when implementing a JIT quality management program. An expected counting measure for all JIT firms should be defects per million units, parts, or items.

Other items that can directly or indirectly measure defective work efforts include counting the number of line stoppages, the total time required for line stoppages, unitary scrap work, material wastage, and rework units. When collected at the work center level of the operation, this information can identify specific departments or workers whose work performance may need correcting. These counts can help identify quality control problems in how workers are doing their jobs or problem areas in an operation where work assignments may need to be rebalanced. Reworked units and number of counted items sent back to vendors during a period of time can be used to assess a vendor's quality performance.

Cost Methods

The cost of quality nonconformity can be divided into four types of costs: preventive costs, appraisal costs, internal failure costs, and external failure costs. **Preventive costs** involve all the costs for efforts to prevent nonconformity. This would include the administrative and training costs for starting and supporting a JIT quality management program. It would also include all the downtime caused by the implementation of the JIT principles. In the long run, these costs will decrease as problems causing nonconformity are found and eliminated. Measuring the decrease in these costs can provide some indication of the successful performance of implementing JIT quality management principles. **Appraisal costs** include the inspection costs used to appraise product quality. Since workers will be performing much of the quality control inspector's role, a drop in the staff function will probably be offset by an increase in workers to perform the added quality control duties. A drop in these costs is beneficial in that a staff function is viewed as "waste adding" rather than "value adding" to the product. **Internal failure costs** include scrap costs, rework costs, and downtime costs. These are all direct costs that should be reduced substantially over time as JIT principles are implemented. **External failure costs** are the costs incurred in dealing with customers or government agencies external to the organization. Under the JIT principles, the costs of warranties, customer suits for product failure, and governmental fines for product failure should all be reduced over time.

Whatever method is selected to evaluate JIT quality management performance, it is important to realize that improvement will take time. Making quality a habit requires behavioral changes, and changing behaviors will take an extended period of time. While the adoption of JIT principles tends not to cause a short-term increase in countable defects, it can typically cause a short-term increase in quality costs. Fortunately, research has shown that over the longer term, both countable defects and their costs will decrease in operations that adopt JIT quality management principles [8].

Integration Strategy for Total Quality Management

To successfully implement a total quality control (TQC) program, employees must be motivated to accept the change the program will entail. Some preliminaries begin motivating employees to accept a TQC program should include: (1) preparing shop floor supervisors to permit subordinates to participate in management activities of planning, directing, and giving orders; (2) assuring employees that the TQC program is a long-term program and not just another management fad; (3) tying reward systems to quality rather than production in units; and (4) communicating the connection between value added activities and job security [9]. Employees have to know that all levels of an organization's management are committed to a TQC program and that they will encourage and reward efforts to make the program a success. With these preliminaries accomplished, a multi-phased approach to TQC can begin.

One implementation strategy for a TQC program includes the following four phases [9]:

1. Awareness The organization should try to establish a common vocabulary for TQC and the need for continuous improvement in the program. This phase can be accomplished by conducting training sessions on product quality, methodology, and general quality control issues. The objective of this phase is to raise the awareness of the importance of quality in all aspects of an employee's job. Management must show a commitment before they can expect to receive a commitment from subordinates.

2. Stand-alone Projects In this phase the organization should designate specific areas or departments where select TQC projects can be implemented. The idea is to show employees what a TQC program can accomplish on a limited basis, so the program's organization-wide TQC program will have successful examples to act as motivators. If employees know that TQC has helped improve the operation in one department, they will be more willing to try it and make it work in their own department. Initial TQC successes can even start some healthy competition between departments to see whose improvements will be the greatest.

3. Integration of JIT Systems The successes in Phase 2 should be brought together into a system-wide restructuring of the operation. The restructuring should be guided by the production department strategic objectives in product quality. Out of this phase should come clear guidelines on what will define a successful TQC program, and an integrated TQC system that states each participant's TQC job-related activities. In addition this phase must determine what management resources will be required to achieve the implementation of a complete TQC program. One tactic to accomplish this phase might be to use quality control circles to act as integration teams. These teams would consist of people from a variety of different areas in the production facility, or throughout the organization. Each team member would represent a different area and bring to the group an area's integration needs and problems. The quality control groups can also help identify solutions to ease the implementation process for management.

4. Making TQC a Way of Life   Making total quality control a habit is a necessary goal that must be a part of the final phase of an implementation plan for TQC. The entire organization's culture and operating systems must be structured for consistency with JIT philosophy and principles of TQC's continuous improvement. This phase requires the establishment of a "quality culture" that embraces all aspects of the organization. Employees must be given encouragement and support for exercising and influencing the direction of the production systems that impact the TQC job-related activities. Effort must also be given in this step to establish long-term goals that are related to strategic organizational goals.

These four phases of implementing a TQC program are very general because their application will vary substantially by the nature of the individual organization, products, and quality objectives. What is constant in the implementation of a TQC program is the need for the JIT quality management principles to be implemented and integrated throughout the entire organization.

Actual Case Study

The Bytex Corporation, based in Southborough, Massachusetts, implemented a TQC quality culture with substantial results [10]. Bytex manufactures network control systems and electronic matrix switches for sophisticated communications users. Their customers include the largest corporations throughout the world.

To implement their TQC program they started with their own people. They structured interdisciplinary teams to study and make recommendations on improving production processes and product quality. The teams identified problems, and suggested ways of integrating JIT inventory, production planning, and quality control methods. They found that their employees, when given the opportunity and encouraged to make suggestions, made significant contributions to the success of the TQC program. The teams identified and eliminated non-value-added activities in production processes. Then inventory reductions were implemented. Once the operation found they could live with the JIT reductions in inventory, they adopted a pull type kanban system to help in production scheduling. These changes led to suggestions for redesigning products, changing supplier contracts to service the JIT operation, and the use of quality control charts that were placed for everybody to see improvements in product quality.

The outcome of these changes toward JIT and TQC operations resulted in a switch whose mean time between failures is 28.5 years (a record for the company, and one of best in the industry). In addition, the Bytex Corporation reports that in the past few years under the JIT and TQC programs they have: (1) reduced total cycle time by 60 percent, (2) reduced inventory by 43 percent (from $6.3 million to $3.5 million), (3) reduced final assembly time by 52 percent, and (4) reduced floor space by more than 30 percent. Reports also confirm that the productivity cycle of TQC has benefited Bytex as their 1989 sales increased by 15 percent over sales from 1988 with a record level of profitability. They are considered one of the fastest growing firms in their industry.

Summary

In this chapter we have examined JIT quality management principles. Consistent with these principles the management of a JIT operation should use process control charting to monitor and motivate workers to achieve product quality perfection. JIT managers should keep the importance of product quality highly visible and strictly enforced, and should seek a continual program of improvement. JIT managers should not only give everybody quality responsibilities, but match that responsibility with the authority to share in the control of product quality. During the expected 100 percent quality inspections performed in a JIT operation, occasional defects will be found. When defects are found they should be corrected by those workers who contribute in generating them. Workers should also be required to perform routine maintenance on equipment and housecleaning duties in their work centers. Managers should be required to maintain long-term quality control records to help instill in workers the habit of quality and the need for product perfection. When these JIT quality management principles are combined with the previously stated JIT principles in the prior chapters, they provide a set of guidelines that will lead to what experts call manufacturing excellence or world class manufacturing [11].

This chapter completes Part I of this book, "An Introduction to JIT Principles." The purpose of Part I is to provide a common basis of JIT principles and methods. Before starting Part II, students may wish to build some practical experience with using JIT principles by running the simulation experiments in Part III. The "JIT Simulation Game" in Part III is designed to provide students with some hands-on practice at using some of the JIT principles of Part I. The application of JIT principles will provide an added learning dimension helpful in dealing with the advanced topics in Part II. In Part II, "Integrating JIT," a series of three chapters is presented in which each deals with a different area where JIT principles are being applied. Selected for their current topical nature, these three chapters describe areas of business operations that represent recent or newly open areas of opportunity for JIT application in the 1990s. The focus of the chapters in Part II is to help explain how JIT can be integrated into these relatively new applications in business, and offer strategies for implementation.

Important Terminology

Appraisal costs the cost of quality control inspection to appraise product quality.

Attribute quality characteristics product features that are used to evaluate quality in a product. Attributes tend to be characteristics that are counted as opposed to being measured.

c-chart a process control chart used to measure quality characteristics, expressed as attributes, that deal with countable defects or flaws where an infinite number of defects can exist in a unit or in a lot of units.

External failure costs the quality control costs incurred in dealing with customers or government agencies external to an organization.

Foolproof devices automated systems used to monitor and control quality in a production process.

Grand mean the average of all sample averages. Used as an estimate for a population average and as a quality standard for the $\bar{x}$ process control chart.

Industrial engineers engineers whose responsibility it is to improve work methods by designing appropriate tools or layouts, and to conduct work studies to determine the most efficient ways of performing work tasks.

Internal failure costs costs to the manufacturer of failure to meet quality standards, including scrap costs, rework costs, and downtime costs.

Job enlargement a management practice of expanding a worker's job tasks to permit a greater variety in work activities. The idea of this practice is to satisfy a worker's need for variety in work and decrease job boredom, thus causing improved morale and increased productivity.

Lower control limit (LCL) lower boundary on a process control chart.

Nonconformity a term used to describe a defect in a process or product.

Normal probability distribution a probability distribution that has a bell-shaped dispersion about its mean value, used to estimate population probabilities based on sample data.

p-chart a process control chart used to measure quality attributes. These *p* quality measures deal with the proportion found defective in a two-outcome decision of defective or not defective for a finite number of inspections on a single item or in a lot.

Preventive costs quality control costs budgeted for efforts to prevent defects.

Process control a management activity of controlling production processes to achieve a desired output.

Process control charts a set of graphic aids used by management to measure and depict quality control efforts in controlling production processes to achieve a desired output. Also called "statistical quality control" (SQC) charts.

Productivity cycling process a logic structure used to explain how JIT can lead to market dominance in any industry (originally presented in Chapter 1).

Quality a concept of conformance to a predefined set of requirements.

Quality audit a management activity to determine if participants in a JIT quality management program are seeking and achieving their quality control goals.

Quality control a management function of verifying conformance to product requirements.

Quality control circle (QCC) a small decision-making group of workers, engineers, and managers, whose purpose it is to solve problems related to product or process quality.

Quality control inspectors staff personnel who help to carry out the quality control management function of verifying conformance to product requirements.

Quality standard a true or population average value measuring the outgoing quality of a production process or product, and used in the construction of process control charts. It can be a calculated value based on samples taken to construct the process control chart, or it can be an assigned value the production process should seek as a standard of quality.

R-chart a process control chart used to measure quality characteristics whose quality standard is expressed as an average variation statistic of range, used to examine quality measures that are continuous in nature.

Small group improvement activities (SGIAs) a term used at Toyota, Inc. of Japan as their name for quality control circles.

Standard deviation a measure of variation used to divide the area under a normal probability distribution into standardized units of dispersion.

Statistical confidence interval theory a field in the subject of statistics devoted to the study of variation and its quantification as a confidence interval about an assumed population average value.

Total quality control a concept involving all aspects of a company in product quality, from product conception through work-in-progress to the delivery of the product. Its goal is to eliminate all contributors of defects in the production process.

Upper control limit (UCL) upper boundary on a process control chart.

Value engineers engineers whose responsibility is chiefly to determine the product materials or designs that will achieve added value to a product at the least cost to the manufacturer.

Variable quality characteristics product features that are used to measure and represent quality in a product; usually those characteristics that can be measured on a continuous scale as opposed to being counted or categorized.

$\bar{x}$-chart a process control chart used to measure quality characteristics that vary on a continuous basis and that can be expressed as measurable lengths, weights, heights, etc.

Discussion Questions

1. What are the JIT quality management principles? Explain at least one tactic useful in implementing each principle.
2. How are the JIT quality management principles different from the JIT production management principles from Chapter 3?
3. How does giving more work responsibility to workers help to improve product quality? How does giving more authority help?
4. How does a JIT operation ensure a 100 percent inspection of every product?
5. How does improving product or process quality help to achieve the benefits of the productivity cycling process?
6. How does a JIT operation use process control charts? How is this use different from their monitoring use in the past?
7. Why are $\bar{x}$- and R-charts usually used together to monitor process quality?
8. When would we use a c-chart instead of a p-chart?
9. What is a QCC, and how can it improve quality in a JIT operation?
10. What measures might we use to evaluate a JIT operation's progress in improving product quality?

Problems

1. A videotape player manufacturer would like to establish a process control chart system to monitor quality in a production system. The tape players manufactured in the production system are inspected by a quality control inspector who tests the finished product by playing each videotape player. If the unit plays a tape, it passes the inspection; if the unit fails to play the tape, the unit fails the inspection. What type of process control chart should be used to monitor quality in this situation? Explain your answer.

2. A compact disc manufacturer would like to establish a process control chart system to monitor quality in a production system. The CD manufacturer has finished CDs inspected by a quality control inspector who examines the finished product. The examination involves recording the number of scratches observed on the disc. What process control chart should be used to monitor quality in this situation? Explain your answer.

3. A small business manufactures pre-cut sticking tape for wrapping gifts in retail stores. Management wants to establish a process control system that will ensure the tape is cut consistently by the cutting machine. The three tape lengths made by the cutting machine are 20 inches, 35 inches, and 40 inches. Cutting longer pieces of tape will waste the tape resources of the business. Cutting shorter pieces will cause scrap, since the cutting must be exact, and customers will reject the product for poor quality. If the company is interested in controlling the length cutting capabilities of the cutting machine, what process control chart or charts should be used in this situation? Explain your answer.

4. A pencil manufacturer wants to establish a process control system that will ensure that the cutting of each pencil is held constant at the desired 7-inch length. The firm has three pencil cutting machines (Machines A, B, and C). A sample of 100 pencils cut by each machine is collected each day for process control purposes. If the company is interested in controlling the length cutting capabilities of the machines, what process control chart or charts should be used in this situation? Explain your answer. Will separate control charts be needed for each machine? Would such detailed charting be of any benefit in locating quality control problems? Explain your answer.

5. A company wants to establish a process control chart system to monitor its manufacturing operation. The company manufactures finishing nails for hardware retail stores. They have decided on $\bar{x}$- and R-charts and have found the grand average of the samples of three-inch nails to be three inches. The average range turned out to be .5 inches on a sample size of 20 nails. What are the UCL and LCL values for the $\bar{x}$-chart? What are the R-chart UCL and LCL values?

6. A local brewery produces a special-order beer product in bottles of 12 ounces. They want to establish $\bar{x}$- and R-charts to monitor the filling machine system. On a sample of 10 bottles of beer each day for 20 days, the grand average was 12 ounces with an average range of 1 ounce. What are the UCL and LCL values for this $\bar{x}$-chart? What are the R-chart UCL and LCL values? How would we use these process control charts in a JIT operation? Explain.

7. A company wants to establish a process control chart system to monitor its dried milk packaging process. The company packages 32-ounce boxes of dried milk. They have decided to establish $\bar{x}$- and R-charts to monitor the quality of the packaging process. The charts are to be based on the following sample data:

Sample Number	Observed Ounces of Dried Milk Packaged in Each Box
1	32, 35, 35, 38, 40
2	32, 29, 32, 34, 33
3	34, 33, 33, 36, 34
4	33, 32, 35, 34, 33

Each of the four samples was obtained by weighing each of five boxes' dried milk contents. The sample number represents consecutive days on which the samples of five boxes are taken from the packaging process. What are the $\bar{x}$- and R-charts' each upper and lower control limits for this company's product? Draw and plot the sample averages and ranges on each of the two charts. Suppose we implement JIT quality control principles and allow two months to pass. If we take another day's sample of five boxes, and it turns out contents are 32, 33, 32, 32, and 32, what would you conclude about the JIT implementation? Show your work by plotting the revised values on both process control charts.

8. A company wants to establish a process control chart system to monitor its candy filling process. The company fills 16-ounce boxes of candy. They have decided to establish $\bar{x}$- and R-charts to monitor the quality of their filling process. The charts are to be based on the following sample data:

Sample Number	Observed Ounces of Candy in Each Box
1	16, 15, 15, 15, 16, 15
2	12, 14, 17, 14, 13, 13
3	14, 12, 13, 16, 34, 14
4	12, 17, 19, 12, 13, 11
5	14, 15, 15, 12, 10, 13
6	16, 16, 14, 14, 16, 16
7	14, 16, 13, 13, 14, 16
8	16, 15, 16, 12, 13, 17

Each of the eight samples was obtained by weighing the contents of six boxes of candy. The sample number represents consecutive days on which the samples of six boxes are taken from the filling process. What are the $\bar{x}$- and R-charts' upper and lower control limits for this company's product? Draw and plot the sample averages and ranges on each of the two charts. Suppose we implement JIT quality control principles. We then take another two days' samples of five boxes each. On Day 9 the candy contents of the six boxes turns out to be 15, 17, 18, 16, 16, and 18. On Day 10 the candy contents of the six boxes turns out to be 16, 19, 19, 16, 17, and 16. What would you conclude about the JIT implementation? Show your work by plotting the appropriate revised values on both process control charts.

9. A microcomputer manufacturing company assembles complete microcomputer systems. Once completed, each unit undergoes a system inspection involving a series of ten individual tests with two possible outcomes for each test: success or failure. Based on a sample of eight microcomputers, the counted number of tests failed for each computer is 3, 5, 6, 3, 1, 1, 2, and 3. What process control chart should be used to monitor quality in this situation? Explain your choice. What is the quality standard of this system? What are the resulting UCL and LCL for this system? Draw the chart and plot the quality control data. Suppose the company now institutes a JIT quality management program. A second sample of eight microcomputers is selected, and the counted number of tests failed are 1, 2, 3, 5, 1, 2, 3, and 1. What are the revised quality standard, UCL, and LCL? Plot the new samples on the revised process control chart. Has the adoption of the JIT principles improved the quality of the microcomputers? Explain your answer.

10. A skateboard manufacturing company assembles skateboards for retail stores. Each skateboard undergoes an inspection for flaws in finish and working parts (i.e., wheels, etc.). Based on a sample of 15 skateboards, the counted number of flaws observed was 1, 3, 7, 2, 4, 4, 4, 6, 3, 8, 6, 10, 7, 8, and 5. What process control chart should be used to monitor quality in this situation? Explain your choice. What is the quality standard for this system? What are the resulting UCL and LCL for this system? Draw the chart and plot the quality control data. Suppose the company now institutes a JIT quality management program. A second sample of ten skateboards is selected and the flaws counted are 2, 2, 2, 1, 8, 11, 3, 4, 7, and 1. What are the revised quality standard, UCL, and LCL? Plot the new samples on the revised process control chart. Has the adoption of the JIT principles improved the quality of the skateboards? Explain your answer.

References

[1]   J. B. Dilworth, *Operations Management*, McGraw-Hill, New York, NY, 1992, chapters 17 and 18.

[2]   R. G. Murdick, B. Render, and R. S. Russell, *Service Operations Management*, Allyn and Bacon, Boston, MA, 1990, chapters 14 and 14s.

[3]   W. J. Stevenson, *Production/Operations Management*, 3rd ed., Irwin, Homewood, IL, 1990, chapter 16.

[4]   R. J. Schonberger and E. M. Knod, *Operations Management*, 4th ed., Business Publications, Plano, TX, 1991, chapters 4 and 15.

[5]   R. J. Schonberger, *Japanese Manufacturing Techniques: Nine Hidden Lessons in Simplicity*, The Free Press, New York, NY, 1982.

[6]   R. N. Kackar, "Taguchi's Quality Philosophy: Analysis and Commentary," *Quality Progress*, December 1986, pp. 21–29.

[7]   D. Elmuti, "Quality Control Circles in Saudi Arabia: A Case Study," *Production and Inventory Management*, Vol. 30, No. 4, 1989, pp. 52–55.

[8]   G. Hohner, "JIT/TQC: Integrating Product Design with Shop Floor Effectiveness," *Industrial Engineering*, Vol. 20, No. 9, September 1988, pp. 42–47.

[9]   J. G. Bonito, "Motivating Employees for Continuous Improvement Efforts," *Production and Inventory Management Review with APICS News*, Vol. 10, No. 6, 1990, pp. 24–26. (Also in Nos. 7 and 8)

[10] B. Dutton, "Switching to Quality Excellence," *Manufacturing Systems*, Vol. 8, No. 3, 1990, pp. 51–53.

[11] R. J. Schonberger, *World Class Manufacturing: The Lessons of Simplicity Applied*, The Free Press, New York, NY, 1986.

CASE 4-1

A Doll of JIT Quality

Dara Toy Company (DTC) is an established toy manufacturer located in Cypress Gardens, Florida. Its current product line consists of ten different toy products, which they market through their own retail stores and franchised stores located throughout the country. The perceived high quality of all of their toy products during the 1980s made their name one of the most respected in the industry. They were able to sell every unit of each product they manufactured within a short time from delivery to retail stores. Many competitive toy manufacturers sought to emulate DTC product quality, but few could equal it during the 1980s.

During the early 1990s, many of the toy manufacturing companies started adopting JIT quality management practices. DTC actually started seeing sales drop for one of their most promising products, called the "sewage patch doll." Marketing research efforts quickly revealed that customers felt the doll was of a very poor quality. Many customers reported arms, legs, and head component parts failing to move in newly purchased dolls. Many of the retail stores started sending defective dolls back to DTC for credit and some of the franchised stores threatened to stop selling the product altogether.

A variety of problems surfaced at the DTC production facility that caused the poor quality in the doll. The sewage patch doll was a hand-assembled product whose delicate parts required a skilled human touch for successful assembly. Unfortunately, the nature of continually unsatisfied customer demand motivated DTC supervisory personnel to stress production quota goals over product quality goals. Many of the older, highly skilled workers became tired of the pushy work environment, and so took early retirement. To make matters worse, the skilled workers were replaced with young and inexperienced workers who had few work skills and were given little production process training.

DTC management realized their doll product quality had fallen to a serious level, and this was a threat to their organization's quality image and long-term success. To deal with the problem, DTC management decided to adopt JIT quality management principles. They took the time to retrain their employees and managers in JIT quality principles. In addition, they felt it necessary to embrace the JIT production management layout design of a flow shop to improve production flow. In Exhibit 4-1 a layout of the flow shop production facility is presented along with a list of the production and quality inspection activities required of workers in each work area.

EXHIBIT 4-1 • *Facility Layout and Work Center Work Activities*

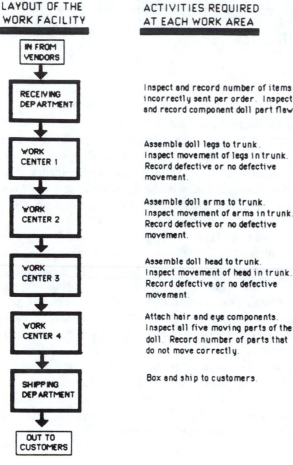

LAYOUT OF THE
WORK FACILITY

ACTIVITIES REQUIRED
AT EACH WORK AREA

IN FROM VENDORS

RECEIVING DEPARTMENT

Inspect and record number of items
incorrectly sent per order. Inspect
and record component doll part flaws.

WORK CENTER 1

Assemble doll legs to trunk.
Inspect movement of legs in trunk.
Record defective or no defective
movement.

WORK CENTER 2

Assemble doll arms to trunk.
Inspect movement of arms in trunk.
Record defective or no defective
movement.

WORK CENTER 3

Assemble doll head to trunk.
Inspect movement of head in trunk.
Record defective or no defective
movement.

WORK CENTER 4

Attach hair and eye components.
Inspect all five moving parts of the
doll. Record number of parts that
do not move correctly.

SHIPPING DEPARTMENT

Box and ship to customers.

OUT TO CUSTOMERS

Case Questions

1. In which work areas would $\bar{x}$- and R-charts be useful in monitoring quality?
    What type of data would have to be collected to prepare these charts?

2. In which work areas would a p-chart be useful in monitoring quality? What
    type of data would have to be collected to prepare this chart?

3. In which work areas would a c-chart be useful in monitoring quality? What type of data would have to be collected to prepare this chart?

4. What are the steps necessary to use the control charts in this facility? Explain in detail how you would implement the charts to achieve the JIT quality management principles and their benefits.

CASE 4-2

It's All a Question of JIT Control

Few things can get an army mad enough to sue a supplier. One area of notable exception involves providing food. The U.S. Army is planning to sue the Good Eats Company (GEC) of Fairfax, Virginia, over just such a matter. The U.S. Army is currently claiming that GEC is at breach of contract with the food product they have been selling the Army for several months.

The problem that is leading to the suit started some time ago when the U.S. Army signed the contract with GEC. GEC agreed to provide the U.S. Army with 12-ounce cans of "span" (a food product that bears some resemblance, but cannot be confused with ham). The contract stipulates that a lot size of 10,000 cans would be sent to the U.S. Army each week. The U.S. Army would then randomly sample the lot by taking 10 cans of span and compare the sample mean weight with the contracted 12-ounce weight. If the sample mean weight is more than ± 0.5 ounces different from the quality standard of a 12-ounce can weight, that week's lot of 10,000 cans could be returned to GEC for credit or replacement. The contract further stipulates that if GEC consistently provides unacceptable cans of span, the contract between GEC and U.S. Army would be considered null and void.

EXHIBIT 4-2 • *Ten Days of Sample (n=10) Weights of Span Cans*

Week Number	Span Can Weights in Ounces
1	13.0, 12.0, 13.5, 12.0, 12.07, 11.95, 11.9, 11.7, 12.06, 12.3
2	12.1, 13.03, 12.0, 12.5, 12.1, 12.07, 13.0, 11.9, 11.87, 12.5
3	13.3, 11.92, 12.1, 12.0, 12.5, 12.0, 12.07, 13.95, 11.9, 11.8
4	12.0, 12.3, 12.0, 12.03, 12.0, 12.51, 12.0, 12.07, 11.8, 11.9
5	13.97, 12.0, 12.3, 12.62, 12.0, 12.0, 13.8, 13.0, 12.07, 11.9
6	12.9, 13.8, 12.06, 13.3, 12.8, 12.13, 12.0, 12.5, 12.09, 12.0
7	13.7, 12.9, 12.8, 12.06, 12.3, 12.02, 12.3, 12.09, 11.5, 12.0
8	12.07, 12.9, 11.9, 11.8, 12.0, 12.3, 12.4, 12.0, 12.8, 12.5
9	12.9, 13.0, 13.95, 11.9, 10.8, 14.2, 12.3, 13.0, 13.6, 13.6
10	14.5, 10.0, 13.0, 13.9, 12.9, 13.8, 14.0, 12.3, 12.0, 13.9

EXHIBIT 4-3 • *Ten Days of Sample (n=10) Weights of Span Cans*

Week Number	Span Can Weights in Ounces
11	11.0, 11.8, 11.5, 12.0, 12.4, 13.9, 13.4, 14.7, 14.6, 14.3
12	11.1, 11.03, 12.0, 12.5, 13.1, 13.07, 13.0, 13.9, 13.8, 14.4
13	11.5, 11.92, 12.1, 12.0, 12.5, 13.8, 13.07, 13.95, 13.9, 13.8
14	11.4, 12.3, 12.0, 12.03, 12.0, 12.51, 13.0, 13.07, 13.8, 13.9
15	12.9, 12.3, 12.3, 12.0, 12.0, 12.0, 12.8, 13.0, 13.07, 13.9
16	12.3, 12.8, 12.8, 12.3, 12.8, 12.13, 12.0, 12.5, 13.09, 13.0
17	12.2, 12.9, 12.0, 12.5, 12.3, 12.02, 12.3, 12.09, 12.5, 13.0
18	12.0, 12.4, 11.9, 12.8, 12.0, 12.3, 12.4, 12.0, 12.8, 12.5
19	12.0, 12.0, 12.5, 12.9, 12.4, 12.2, 12.3, 12.0, 12.6, 12.6
20	12.1, 12.0, 12.0, 12.3, 12.6, 12.8, 12.0, 12.3, 12.0, 12.3

In the first eight weeks of the contract the 80,000 cans (i.e., 10,000 per week) were delivered by GEC and judged acceptable by the U.S. Army. Both GEC and the U.S. Army were very happy with their contractual arrangement, though Army inspectors did notice some variation from the desired 12-ounce can weight goal. In the 9th week, the U.S. Army rejected the lot based on a sample mean of 12.9 ounces. In the 10th week the U.S. Army also rejected the lot based on a sample mean of 13.03 ounces. The actual sampled can weights for the first ten weeks are presented in Exhibit 4-2. GEC was surprised that the U.S. Army rejected this lot so soon after rejecting the last one. While GEC did not have any process control system in place, they did check their manufacturing equipment and found it to be performing as expected. They asked some workers to pull a sample batch of span cans off the line. Based on this sample, they found the mean weight to be exactly 12 ounces.

The rejected lots in the 9th and 10th weeks were very costly to reprocess by GEC. While the two reprocessed lots were eventually sold to other customers, GEC did not make any profit on either lot because of the extra delivery costs. The U.S. Army, on the other hand, had incurred inspection and sampling costs, as well as rush delivery charges to obtain replacement span to feed the troops. They were not pleased with the poor service and poor quality from GEC.

The base acquisition officer for the U.S. Army called GEC and informed the management that they would suspend the contract with GEC for ten weeks. The ten-week suspension was to allow GEC to change their production processes to achieve the U.S. Army quality tolerance expectations stated in the contract. It was not expected that GEC's span can weight would always meet the U.S. Army's goal, since other factors (such as sampling error) could be causing the deviation, rather than flaws in the production process itself. The U.S. Army wanted to have some specific proof that GEC was making some progress toward falling within ± 0.5 ounces of the 12-ounce weight goal. If not, GEC's contract with the U.S. Army would be considered null and void. Even more critical, the U.S. Army would sue GEC for replacement costs they incurred in finding another supplier.

GEC really wants the U.S. Army contract, and realizes they should do a better job at filling the span cans. They agree to a trial period and the proof of success criteria for continued business. They also decide to adopt JIT quality management principles and to install an $\bar{x}$-chart process control system as a means of monitoring quality. They hope the results of the ongoing implementation of JIT principles will be reflected in the process control chart with an improvement in span can content weight. GEC decides to use the sample data collected in Exhibit 4-2 to develop the chart. The sample data collected during the ten-week trial period is presented in Exhibit 4-3.

Case Questions

1. What are the UCL and LCL if we arbitrarily set the quality standard at 12 ounces and use the sample data in Exhibit 4-2? Draw the $\bar{x}$-chart for this question and plot the ten weeks of sample averages.

2. Take the sample for Week 11 from Exhibit 4-3, compute its average and add it to the computations to revise the UCL and LCL in the $\bar{x}$-chart in Question 1. Show how the impact of the new data has changed the control lines by drawing them on the $\bar{x}$-chart from Question 1. Has quality improved or not? What would you expect in the first week during a changeover to JIT quality principles? Explain.

3. Take the sample for Week 12 from Exhibit 4-3, compute its average and add it to the computations to revise the UCL and LCL in the $\bar{x}$-chart in Question 1. Show how the impact of the new data has changed the control lines by drawing them on the $\bar{x}$-chart from Questions 1 and 2. Has quality improved or not? What would you expect in the second week during a changeover to JIT quality principles? Explain. Repeat this process for the remaining eight weeks, plotting the new UCLs and LCLs as you go. How would you interpret the impact of JIT quality principles on the filling process based on the resulting revised chart?

4. What should GEC do for a defense against declaring the contract null and void? Explain your defense. How can the R-chart be used in the defense? Show your work.

PART TWO

Integrating JIT

CHAPTER FIVE

Integrating JIT with Computer Integrated Manufacturing (CIM) Systems

Chapter Outline

Introduction

Overview of Computer Integrated Manufacturing

Evolution Toward a Computer Integrated Manufacturing System

Composition of a CIM System

Benefits of CIM and Their Relationship with JIT

Integrating JIT and CIM

A CIM Software System That Supports JIT

Common Elements in CIM-JIT Systems

A Methodology for JIT Software Selection

Learning Objectives

After completing this chapter, you should be able to:

1. Explain what a computer integrated manufacturing or CIM system is composed of and how it benefits users.
2. Describe the basic modules of a JIT software system and explain how they support JIT principles.
3. Explain how CIM and JIT can be integrated.
4. Explain some of the benefits of a combined CIM-JIT system.
5. Explain a methodology that can be used to select the "best" JIT software system.

Introduction

A **computer integrated manufacturing (CIM)** system is the application of computer hardware and software to achieve an integrated manufacturing system that is controlled in part or wholly by a computer system. (The composition and components of CIM will be discussed in the next section.) A CIM is a manufacturing system that operates under the control of a computer and within the limitations of the automated equipment that composes the system. Many people view CIM systems as inflexible, and have actually designed statically controlled CIM systems to be fairly inflexible in what they can produce and in the way they produce them. In these systems component part inventory is fed to computer controlled equipment that performs a specific set of instructions. No opportunity for improvement or change that might benefit the system is possible [1,2,3,4].

This limited perception of what CIM is, is typical of the late 1970s when CIM was trying to emerge from the **islands of automation** that had little or no computer integration, but a great deal of highly **dedicated automation** [5]. Foreign competition during the 1980s taught U.S. manufacturers a hard lesson in the need for improving flexibility in automated systems. The highly flexible systems, like JIT, used by Japanese companies captured U.S. markets by providing responsive changes in products and services that U.S. companies could not meet with their static CIMs of the period [6]. In response to this need, a number of highly automated **flexible manufacturing systems (FMS)** became popular in the 1980s [7]. Research on how to implement flexible manufacturing systems was also quite popular during the 1980s [8, 9]. These systems met with only partial success, because they lacked much needed computer integration. They are considered an important part of what makes up a modern CIM. Computer integration could only be achieved by improved computer software systems. A number of computer software systems were developed during the 1980s that made CIM a reality.

The need to further move toward building flexibility into CIM, coupled with the timely advancement of JIT during the 1980s, led product experts to call for the integration of CIM and JIT into a common system [10, 11]. During the same time, a number of computer companies started developing CIM software that would support JIT operations within a CIM system environment. The Hewlett Packard software system, "HP Manufacturing Management II", and the IBM software system, "Manufacturing Accounting and Production Information Control System/Data Base" (MAPICS/DB) are just two of the more commonly used CIM systems that support JIT production activities [12, 13]. Recent research on manufacturing operations that have been trying to achieve some limited integration of these two systems has reported joint benefits that are unattainable by either system independently [14].

The purpose of this chapter is to examine the integration of JIT with CIM. Specifically, this chapter will explain how JIT and CIM can work together to accomplish the joint production benefits of JIT within a CIM work environment. To accomplish this objective, we will begin with an overview of what a CIM system consists of and how it operates in a modern manufacturing operation. We will then focus on how JIT is integrated into a CIM system with a discussion on a current software system that supports the combined CIM-JIT manufacturing system. A method for selecting a JIT software system is also presented.

Overview of Computer Integrated Manufacturing

Evolution Toward a Computer Integrated Manufacturing System

Many manufacturing organizations are moving toward a CIM operation in the evolutionary direction depicted in Figure 5-1. Today, few organizations have achieved the fully computer integrated system at Stage 4. In general, manufacturers tend to start out with small, manual production operations and move in a step-like progression toward CIM [15]. Like a pyramid, a production system moves from non-computer Stage 1 toward the CIM in Stage 4 of Figure 5-1. Not all organizations, though, want to be fully integrated, even in the 1990s. Small volume operations, businesses offering products requiring a high degree of human skill to produce, capital-poor businesses unable to afford automation, and JIT operations that work well without the need for computers might all be operations where manual methods are the best choice. In general, though, all businesses are becoming more computer-oriented, computer-invested, and computer-

FIGURE 5-1 • *A Production System's Progression toward Computer Integration*

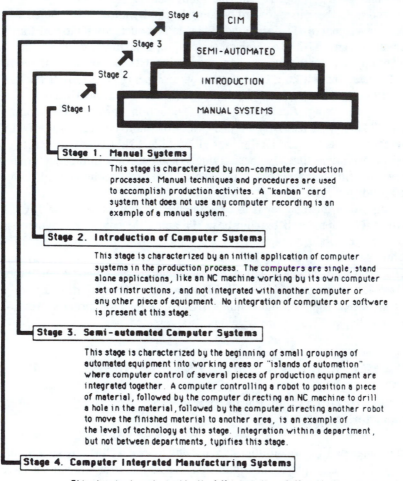

Stage 4 — CIM

Stage 3 — SEMI-AUTOMATED

Stage 2 — INTRODUCTION

Stage 1 — MANUAL SYSTEMS

Stage 1. Manual Systems

This stage is characterized by non-computer production processes. Manual techniques and procedures are used to accomplish production activites. A "kanban" card system that does not use any computer recording is an example of a manual system.

Stage 2. Introduction of Computer Systems

This stage is characterized by an initial application of computer systems in the production process. The computers are single, stand alone applications, like an NC machine working by its own computer set of instructions, and not integrated with another computer or any other piece of equipment. No integration of computers or software is present at this stage.

Stage 3. Semi-automated Computer Systems

This stage is characterized by the beginning of small groupings of automated equipment into working areas or "islands of automation" where computer control of several pieces of production equipment are integrated together. A computer controlling a robot to position a piece of material, followed by the computer directing an NC machine to drill a hole in the material, followed by the computer directing another robot to move the finished material to another area, is an example of the level of technology at this stage. Integration within a department, but not between departments, typifies this stage.

Stage 4. Computer Integrated Manufacturing Systems

This stage is characterized by the full integration of all production processes Every production activity is either controlled by or monitored by the computer system. A CIM is also integrated with the organization's MIS so information on production and the activities of the rest of the organization is shared by all.

integrated. Competition is forcing many organizations to more quickly move toward Stage 4 operation to take advantage of the many benefits of CIM.

The objectives of CIM systems and the resulting benefits that come from their accomplishment are in part what is causing many organizations to seek this type of facility for their operation. The CIM production system objectives include: (1) the need for greater flexibility in production processes to meet the rapidly changing market of the 1990s, (2) timely production information to improve management planning and control activities, and (3) improved product quality through automation. These objectives share common ground with those of JIT discussed in prior chapters. (We will discuss the benefits of CIM and their relation to JIT in a later section of this chapter.)

To understand how CIM and JIT can be combined, we must first understand the CIM system. Let's look at the composition of a typical CIM operation.

Composition of a CIM System

The purpose of this section is to present an overview of what makes up a CIM system's hardware and software. Let's examine the typical CIM operation presented in Figure 5-2. As we can see in Figure 5-2, customer orders are taken and entered into the computer system by marketing personnel using a computer terminal or a microcomputer that is networked to the manufacturing organization's **mainframe computer** (i.e., a large-sized computer). The mainframe computer and its software are often referred to as an organization's **management information system (MIS)** since it integrates information resources throughout the organization. The mainframe computer acts as an initiator to direct the beginning of all of the computerized production activities. The mainframe converts the marketing order into a production order by referencing its **data base** (i.e., an extensive file of information) customer product files for product specifications.

The mini-computer performs as a decentralized work order dispatcher. The mini-computer, using its data base, defines the specific work requirements for each product, on each order, and conveys them to a number of different technologies, including:

1. Individual Work Center Computer Terminals A **computer terminal** is a communication device that is usually hard wired directly to the mini-computer and provides human users a means of interactive communication with the computer system. A terminal is used to retrieve detailed work order instructions, listings of work tasks, and locations of component parts or materials for production from the computer. In some operations **microcomputers** are used in place of computer terminals to provide both a communication role and independent computer processing capability.

2. Robots A **robot** is a mechanical device used to pick up, move, or position materials, or perform production tasks. Some robots are **dedicated machines** with limited capabilities, while other robots may possess their own computer systems capable of **artificial intelligence (AI)** and independent actions.

3. Automated Guided Vehicles (AGV) An **automated guided vehicle** is a motorized pallet mover that may be centrally controlled by a facility's computer or

FIGURE 5-2 • A Computer Integrated Manufacturing (CIM) system

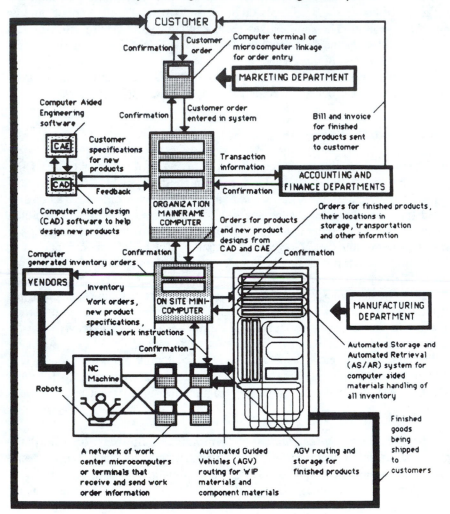

may possess independent computer control. Some AGVs may possess sensory capabilities and a limited artificial intelligence to enhance the problem-solving efforts in material handling (e.g., what to do when faced with a closed door blocking its path). The AGVs permit great flexibility in routing work-in-process. This flexible routing capacity permits them to implement unitary production in the highly automated environment of a CIM since re-routing can be performed for each unit.

4. Numerically Controlled (NC) Machines **Numerically controlled (NC) machines** are computer-based tools that are programmable to perform a limited number of operating instructions. A typical example of an NC machine is an automated drill press that can change drill bits and/or drill a hole in a human- or robot-positioned piece of metal. When NC machines and robots or AGVs are integrated together and programmed to handle a variety of different products, it represents an example of **flexible manufacturing systems (FMS)**.

5. Automated Storage and Automated Retrieval (AS/AR) Systems An **automated storage and automated retrieval system** is an integrated inventory storage and retrieval system that performs material handling and order processing and even helps in transportation loading activities. It is a computer controlled system that can locate inventory, pick items to fill orders, and deliver them to loading docks without the need for human assistance.

6. Electronic Sensors **Electronic sensors** are devices used to monitor production equipment or products. Electronic sensors generate data that are conveyed electronically to computer systems where software is used to determine if adjustments are necessary to equipment or products, and what these adjustments are. These sensors are used to help robots locate and position parts for processing by equipment. These sensors are also an important part of an automated product quality control inspection program as they are used to collect necessary product quality information. Such inspection systems might include **optical scanning equipment** used to read printed characters or **bar codes** for the recognition of characters by their images.

The network of work center computer terminals or microcomputers is usually linked with the mini-computer. The mini-computer may also provide an automatic link with vendors. When orders for products require component parts or materials from vendors, the mini-computer may automatically generate inventory orders for purchasing department personnel to obtain the required parts or materials.

The mini-computers (and mainframe computers) use a variety of manufacturing software to plan, schedule and direct the activities of the computer systems and human resources at their command. Some of the more commonly used methods and techniques include:

1. Computer Aided Design (CAD) Software **Computer aided design** software is used by draftspeople and other creative technicians to create, design and graphically present images of new product ideas. Sometimes called an electronic drafting board, the CAD software eases the conversion of product ideas into graphic two or even three dimensional images for design purposes. The CAD system also provides design specifications from the graphic representations to make physical product modeling efforts easier and more accurate.

2. Computer Aided Manufacturing (CAM) System Software In some operations, CAD systems are connected to **computer aided manufacturing (CAM)** systems that can convert the CAD designs into computer instructions that can be understood by different automated equipment. The combination of CAD/CAM systems is considered the forerunner of CIM systems since they are capable of controlling and directing automated islands of production equipment to produce products.

3. Computer Aided Engineering (CAE) Software **Computer aided engineering** software continues where CAD leaves off. The CAE software takes the computer generated information from the CAD system about a product and allows engineers to conduct mathematical and simulated tests on design characteristics, such as structural integrity or arrow dynamics. CAE software saves redesigning costs by allowing new product designs to be tested and flaws overcome before physical models of the CAD design are produced.

4. Group Technology (GT) As stated in Chapter 3, **group technology** is a method of combining production efforts that seeks to identify the sameness of parts, equipment, or production processes to take advantage of common setup efforts. It is ideal for automated systems that are limited to producing a narrow range of products. GT methods are able to easily adopt automation in place of humans, and take advantage of their high efficiency.

5. Material Requirements Planning (MRP) Software System A **material requirements planning system** uses software to help plan and control the use of dependent demand inventory items and schedule finished product production. This software assumes an infinite amount of production capacity exists to produce the MRP-generated schedule and uses simple logic to complete production schedules. (We will be discussing MRP in Chapter 6.)

6. Optimized Production Technology (OPT) Software An **optimized production technology** system uses software similar to an MRP system, but considers production capacity constraints and schedules finished product production based on advanced mathematical optimization procedures.

Once a work order is received at a work center by a human or robot, the work is performed, and the results of the finished work are entered into the computer system for control and order tracking purposes. For finished manufactured products (or WIP within the manufacturing department), the mini-computer might direct an automated guided vehicle (AGV) to perform material handling activities and move the products to appropriate locations for additional processing or storage.

The mini-computer may also be in control of the automated storage and automated retrieval (AS/AR) area within the production department or in a separate facility. The AS/AR system can be an almost fully automated operation that stores, retrieves, and prepares product orders for shipping. The mini-computer may direct the automated systems, including robots and AGVs, to specific product storage locations so customer orders are picked, packed, and shipped according to the customer's unique specifications.

At the same time that ordered units are being shipped to the customer, the mini-computer processes the customer's bill by confirming units and products shipped, backordering units not shipped, or canceling orders. The information on the order is sent from the mini-computer back to the mainframe computer, which channels it through the finance and accounting departments to perform their transactions before sending invoices to customers. CIM systems greatly expedite accounting and financial transactions, activities that are very time consuming for non-computer integrated operations.

The degree of computer integration of the CIM system in Figure 5-2, or in any other system, is dependent on the amount of work controlled and directed by computers. The more the contribution from human resources, the less the contribution from computers. No organization, as of the early 1990s, has ever achieved a 100 percent fully integrated CIM system. Few organizations will ever want a 100 percent level of integration, but many in the 1990s will move up the pyramid in Figure 5-1 as their manufacturing and other systems become more computer integrated. What will be driving them will be the benefits that CIM has to offer all operations, including those performed under JIT principles.

Benefits of CIM and their Relationship with JIT

In using the CIM presented in Figure 5-2, it would appear that a considerable capital investment is necessary to implement a CIM system, and that the costs of such an investment might prohibit its development. Research on the costs and benefits of modern CIM systems seems to demonstrate that the benefits are definable and can far outweigh cost considerations in most manufacturing operations [16, 17]. While there are many benefits of using a CIM system, three of particular importance to JIT deal with issues of flexibility, timely information, and product quality.

1. CIM Flexibility CIM advanced technologies allow a variety of ways to achieve greater manufacturing flexibility than was previously available, except at extremely high cost. One area where CIM exceeds what JIT can do alone is CIM's capability to equally handle any type of product regardless of its complexity. JIT is viewed as an ideal production system in situations where the sequence of production activity is less complex, and where production lead times are small. As the number of job tasks increases, and as lead times are lengthened, JIT's manual methods tend not to be productive. Very complex products require a sequence of production activities that make it more difficult to implement JIT visual management methods, or improve work methods for jobs that have unchangeably fixed, long lead times. CIM systems, on the other hand, are designed to keep track of complex projects that are not visually understandable, and to plan better the timing of jobs that have long lead times. Other areas of flexibility CIM brings to manufacturing systems include: (1) volume (varying the amount of production during a planning period), (2) parts (being able to handle new parts in a product mix), (3) product mix (changing the proportion of parts that will be added to a product mix), (4) design (allowing engineers to change product designs easily), (5) sequence (allowing for variations in the sequence ordering of parts in WIP), (6) routing (being able to re-route parts to improve worker or machine utilization), (7) materials (being better able to adapt to changes in inventory materials), and (8) integration (being able to take advantage of changes in technology and integrate them into the production process) [16]. These are many of the same benefits of using a non-computer-based JIT production system. The difference between JIT flexibility and CIM flexibility is that JIT not only allows flexibility, but also motivates improvements. A CIM system doesn't seek to motivate improvements, but allows for their implementation. This means that a CIM system supports JIT's need for flexibility and can accommodate JIT's extra process and product improvement benefits. On the issue of flexibility, an integrated CIM-JIT system complements each other's software applications, and allows users to take advantage of both systems' benefits.

   2. CIM Timely Information   CIM's advanced technologies allow for more timely information than possible under previous manufacturing systems. CIM provides more timely information about: (1) customer order processing (i.e., confirmation of the order received from the customer, reliable estimates on its completion, confirmation of the order completion); (2) production planning and control (i.e., keeping track of cost information, customer orders, amounts and location of inventory and materials, amounts and location of WIP, machine usage time, worker center usage time, and generation of periodic status reports on these system elements as compared with predefined standards); and (3) vendors (i.e., automatically generates purchase orders for inventory when required, maintains track records on vendor service quality, generates periodic reports on the status of vendors as compared with predefined standards). These timely information benefits are one of the basic reasons why manufacturers use a CIM production system. Since most of these activities are considered as "collecting" or "recording" activities, they are viewed in a JIT operation as non-value-adding or wasteful activities. Yet in many industries, particularly those with custom products and lower volume production, timely information provides a key competitive advantage for successful operations. Customer order processing time information can make the difference in making a sale or not, particularly where order lead time is the critical factor for a sale. In industries where product cost is critical, CIM planning and control information can provide a more precise cost estimate to help make the sale. By integrating vendor information systems by **electronic data interchange (EDI)** (as discussed in Chapter 2), up-to-date cost and delivery information can also be incorporated into customer order processing to provide more accurate and reliable cost estimates. The difference between JIT and CIM in this type of "timely information" is that JIT does not automatically collect this information and even seeks to discourage its collection because it considers this a waste of time. Recent introduction of optical scanning equipment and bar coding of materials, WIP, and work activities at the work center level of a manufacturing operation has significantly reduced the "waste of time" consideration of collecting this data in a CIM environment. (Remember we discussed the use of bar coding as a means of implementing "kanban" systems for JIT operations in Chapter 3.) A CIM system is equipped not only to collect this information efficiently, but to make it available throughout an organization on a "real-time" basis for timely decision making, which in turn means greater flexibility. Also, as businesses move to provide greater service to their customers, their marketing departments recognize the value in the service of information. Information for the customer has become a significant part of the total product that a company offers its customers. The automated equipment of modern CIMs minimizes possible JIT objections to data collection effort and supports other JIT principles, while it provides useful planning and control information.

3. CIM Product Quality   CIM advanced automation minimizes process and product variation by removing the major cause of variation in a production system: human workers. Robots, AS/AR, and other automated systems replace, but not completely, human resources that are chiefly responsible for production process variation. Modern CIM systems actually reallocate human resources to those activities that people do best. Humans are assigned to tasks that require human intelligence and a higher level of flexibility than an organization can obtain from automated systems. This is the same logic as the JIT production management principle of using automation where practical (from Chapter 3). As suggested by JIT principles, automated systems or foolproof devices should be integrated into manufacturing systems to ensure product quality. A CIM is the ideal system to receive and integrate quality control technology since it is already electronically based. Indeed, in a stand-alone JIT environment, quality control technology generates information limited to the area where the machine is positioned. This limited domain for shared information actually runs counter to another JIT production management principle of improved communication and visual control. An integrated CIM environment, on the other hand, permits the sharing of all data through a common management information system. The integration of the process control equipment and its processing of information will not generate JIT waste, but instead provides a shared basis of data that can improve product quality planning. A CIM system enhances a JIT's use of automated quality control equipment.

In summary, the benefits of CIM complement, support, and can enhance those of JIT. The need for flexibility in both systems has a tendency to make the systems complement one another. The CIM benefit of timely information can enhance a JIT system without major inefficiency due to CIM's technology. The flexibility in CIM to support the dynamic JIT role of improving product and process quality can result in a complementary synergy that experts feel will lead to competitive excellence [18]. Integrating CIM and JIT will result in shared and unique benefits to both systems that cannot be obtained from either system separately [14].

Integrating JIT and CIM

The shared objectives and principles of CIM and JIT have led some experts to view JIT as an implementation strategy for the improvement of production processes, while a fully integrated CIM is viewed as the long-term objective of the implementation strategy [5,10,11]. Their approach can be characterized as a three-step implementation process of: (1) simplification and improvement of products and processes with JIT principles, (2) selective automation, and finally (3) CIM integration. This strategy may be useful for companies just starting up, but what about the majority of manufacturing companies that are currently operating with a system somewhere on the pyramid of CIM development depicted in Figure 5-1? In this more common situation experts feel that JIT is an essential component of modern CIM systems and its principles should be applied fully and completely within any stage of CIM development [19].

Integrating CIM and JIT will depend on where a company starts. For organizations that have achieved some degree of computer integration (Stage 3 or Stage 4 in Figure 5-1), the integration of CIM and JIT must be supported with computer software that defends JIT principles. A company need only obtain and implement the JIT software within the CIM system that is in place. The company also has to adopt the JIT principles embraced by the software system. (Since we have already introduced the JIT principles and implementation strategies in earlier chapters, we will not restate them here.)

For JIT organizations that solely operate using manual methods that embrace JIT principles, the integration of CIM and JIT involves a step-wise process. From a non-computer JIT beginning (the Manual Systems of Stage 1 in Figure 5-1) automation is introduced systematically. Based on JIT principles, automation is introduced in a production system in areas such as process control, material handling, and handling of high-volume repetitive production activities. This moves the JIT operation to Stage 2 (Introduction of Computer Systems). Consistent with JIT production management principles, a JIT operation will use GT cells as a production layout framework. The acquisition and implementation of computers and software to control the production activities of each GT cell moves the JIT operation to the Stage 3 (Semi-automated Computer Systems) level of CIM development. At this level, each separate GT cell might have its own independent computer system controlling just the activities of that particular cell. Both at Stage 3 and finally at Stage 4, the development of CIM is dependent on having computer software capable of interpreting and converting differing computer hardware languages into an integrated language that can be used by a central computer to coordinate production activities. To handle this interpreting job requires special computer languages called **protocol languages.** One of the more significant developers of protocol languages is General Motors, who in the 1980s developed their **Manufacturing Automation Protocal**™ **(MAP)** system of standards used to facilitate communication between computer equipment.

A CIM Software System that Supports JIT

Whether from CIM to JIT or JIT to CIM, a software system is the critical key to combining these two manufacturing approaches. There are a number of competing software systems in the market that provide support for JIT in a CIM environment. The Arthur Andersen & Company MAC-PAC/JIT [24] and the Hewlett Packard HP Manufacturing Management II [12] systems are two such systems. One of the more recent versions of a CIM system supporting JIT with computer software is the IBM system called "Manufacturing Accounting and Production Information Control System/Data Base" (MAPICS/DB) [13]. This system has many of the same features of the MAC-PAC/JIT and HP systems, and is similar in integration to the system depicted in Figure 5-2. The software modules of IBM CIM-JIT system include:

1. *Plant Operations* This module integrates and performs functions in inventory management, production control and costing, production monitoring and control, and purchasing.
2. *Marketing and Physical Distribution* This module integrates and performs functions in order entry and invoicing, sales analysis, and forecasting.
3. *Production Planning* This module integrates and prepares the master production schedule (MPS), material requirements planning (MRP), capacity requirements planning (CRP), and product management activities.
4. *Financial Management and Business Control* This module integrates accounting and finance activities including general ledger, accounts payable, accounts receivable, and payroll transactions. The module also performs a number of financial analyses for planning purposes.
5. *Cross Application Support* This module provides assistance to users for purposes of installation, operation, customizing reports and software functions, and maintenance activities.

Collectively, these modules form the basis of the CIM system that will direct the manufacturing activities of the system and conduct the interrelated marketing, financial, and accounting activities that support manufacturing. To further enhance the MAPICS/DB system, IBM developed a number of optional modules that can be integrated into the system. These modules are particularly important as they offer a tactical means of enhancing the CIM system to support JIT. These optional modules include:

1. Plant Operations Interface This module provides the software communication link between the rest of the CIM system and the shop floor devices like robots, NC machines, and bar code readers. The module allows detailed information such as labor and material usage information to be shared with other areas within the organization. It also offers users the flexibility to run this system on other computers than those of IBM. Of particular importance, it allows greater production flexibility as required in a JIT operation. This module allows flexibility in: (1) accessing relevant communication by using an AI based processor (thus enhancing the JIT concept of facilitating communication), (2) plant layout by allowing changes in devices or networks (which is a JIT means of improving efficiency), and (3) allowing changes in job orders or lot sizes to meet with demand changes (which is a JIT principle).

2. Electronic Data Interchange (EDI) Interface This module brings EDI to order entry and invoicing, purchasing, and accounts payable. EDI brings with it the JIT principle of improved communication and the JIT goal of removing waste by a reduction in order processing time, a reduction in delivery time, improved accuracy in order processing and accounting transactions, and as stated in Chapter 2, a reduction in labor costs.

3. International Support Enhancements This module supports an organization's growth, as expected under a JIT system, by assisting it as it moves into international markets. This module helps in purchasing and order entry by taking into consideration multiple currencies in the processing of foreign invoices and orders. It helps in accounting and financial areas by being able to provide custom information consistent with the governmental guidelines in those foreign operations and translations for U.S. operations.

4. Group Job Support This module is devoted to integrating the office operations with the manufacturing facility activities. This module provides users with a computer terminal and a menu-driven interactive software system offering a number of office features including electronic calendars, access to library services, creation and distribution of inter-office communications, mailing services, and personal directories. These features support the JIT principle of avoiding wasted time in non-value-added activities. (We will be discussing JIT in administrative office management in Chapter 7.)

5. Repetitive Production Management This module is specifically designed to integrate JIT principles into a CIM environment. Recognizing that JIT is a dynamic motivator for change, and CIM systems tend to be fixed in nature, this module is built to accommodate the variety of process and product improvement changes that JIT is expected to contribute to a manufacturing operation. In the area of production scheduling this module supports the uniform production smoothing required in a JIT operation and can also support lot size production using MRP and OPT scheduling software if needed. Its flexibility allows daily scheduling revisions to meet changes in demand and the ability to vary the sequence of production scheduling to accommodate a wide range of product families. One advantage of this CIM system over a non-computer JIT system in production scheduling is that materials, inventory, and capacity information reports can be provided on a daily basis to confirm work order completion. Unlike JIT, the module also performs all accounting and financial transactions on a real-time basis to maintain strict cost control of production. The MAPICS/DB is one of the few CIMs that supports the electronic use of kanban systems. This module gives the CIM the capability to use kanban cards to implement the synchronized demand pull system required for JIT production, while using the computer-based technology to minimize wasted time (as described in Chapter 3).

These modules collectively provide a complete MIS for a manufacturing operation. It is a system that can grow with additional modules to embrace and support every aspect of the organization that currently uses humans or robots.

Common Elements in CIM-JIT Systems

How do the CIM software systems integrate JIT principles to their mutual benefit? There are a few common elements of combined CIM-JIT software systems that are specifically focused on key JIT management principles. They provide a starting place for the implementation of a JIT system within a CIM environment. These common elements include:

1. Involvement of People CIM-JIT systems allow flexibility to use as much or as little human resources as a firm feels it needs. By locating robots in work centers, humans can substitute for automation until technology proves it can do a better job or until automation becomes financially feasible to justify replacing people. Since most of the creative ideas and suggestions for improvement in product quality and production processes originate from workers, it is doubtful that any firm will ever want to replace all of its workers. Even the AI-equipped computer systems don't come up with new ideas on how to improve production flow. These ideas generally come from workers who make the production system flow and are a critical part of JIT. CIM-JIT systems are designed to allow flexibility in implementing the JIT improvement suggestions workers make quickly and without the need for substantial reprogramming.

2. Demand Pull System The workers and management are free in a CIM-JIT system to perform under all of the same JIT production management principles as stated in Chapter 3. The CIM-JIT system allows daily production scheduling to synchronize with customer demand. Workers are allowed to determine production flow, consistent with JIT production management principles. The software scheduling systems are based on the mixed model scheduling method (Chapter 3) and utilize GT cell production setup benefits. Each work center along a GT cell in a CIM-JIT system can be operated by a worker or some type of automation like a robot. In work center cells that are completely automated, the CIM system can reprogram equipment to perform new work tasks far more quickly than human workers can be retrained. This means that a CIM fully automated GT cell can reduce setup time (by reprogramming itself and mechanically repositioning equipment) and help achieve the JIT goal of unitary production more efficiently than by using human-only JIT methods. Of course humans still have far more flexibility than any existing automated system. If workers are operating a work center, their work instructions are sent to them by their computer terminals. When they complete a unit of work, either they can acknowledge their completion manually by typing a code on the terminal or optical scanners can be deployed to confirm a unit is finished, and is being sent along the line. When units of production are finished at various points along a GT cell, the WIP inventory used to complete a unit can be automatically deducted from total inventory by the computer system to record and keep track of inventory transactions on a real-time basis. (This deducting process will be discussed in Chapter 6.)

3. *Total Quality Control (TQC)* CIM-JIT systems support all the JIT quality management principles stated in Chapter 4. Moreover, the CIM portion of the system enhances JIT principles in process control, communication and maintenance activities. The CIM system can easily collect and generate the desired process control charts used in a JIT operation on a real-time basis over computer terminals at the individual work centers. Workers will be able to immediately see the electronically measured impact of their efforts in quality. The CIM system, being integrated, also has the capability of graphically presenting the quality of one worker's efforts to others, and send quality measurements made later by others back up the line to workers. Sharing this information via the integrated computer improves communication and helps make quality efforts more visible as JIT requires. The CIM system can also be structured to keep track of and inform workers about routine maintenance activities, and can be used by maintenance specialists to monitor excessive worker efforts to keep machines working beyond their useful life, when these machines actually should be replaced.

4. *An Integrated JIT Vendor Interface* CIM-JIT systems support all the JIT purchasing and supplier relationship characteristics discussed in Chapter 2. In addition, the CIM system can automatically track a number of criteria used to measure vendor performance. This automatic monitoring and recording activity can improve efficiency in the JIT operation by avoiding the wasted time of recording these much needed purchasing evaluation criteria of the 1990s [20, 21].

5. Support for the Focused Factory CIM-JIT systems are structured to support the synchronized flow shop layout utilizing GT cell configurations and product family scheduling. Workers are free to perform their activities at work centers in the same manner as they would in a completely non-computer system. The CIM system, though, can provide reminders and helpful information for workers via their computer terminals (e.g., detailed work instructions, locations of materials to do job assignments, etc.). Moreover, the CIM can be used to automatically monitor and record labor and equipment usage rates in the performance of job tasks. This usage information can later be used for planning labor and equipment capacity requirements.

In summary, integrating CIM and JIT can be achieved without sacrificing the benefits of either system. Indeed, CIM systems can enhance and help implement JIT principles. The simplicity of JIT systems can operate as well, if not better in the complexity of a CIM system. Together these systems can offer manufacturers a powerful system of dynamic improvement with greater control and planning capabilities than when implementing either system individually.

A Methodology for JIT Software Selection

The key to implementing a CIM-JIT system is having software that supports JIT principles. Since each organization's approach to JIT principles varies as a function of a number of organization variables including type, size, and objectives, great care must be taken in the software selection process. Selecting JIT software that best fits the unique importance attached to JIT principles for each individual organization should be considered in any method used to help select JIT software. JIT inventory principles might be more important for a distribution organization that is predominantly inventory-oriented, and production principles less important. These differences must be considered in selecting the "best fitting" software.

One method that can be used to help in the selection of JIT software is the use of scoring models [16]. A **scoring model** uses a collection of quantitatively evaluated selection criteria to help decision makers reach a decision. Scoring models are judgmental methods based on the expertise of one or more decision makers. This methodology has been used in selecting CIM systems and was introduced in Chapter 2 as a means of selecting JIT vendors [21].

When applied to select a JIT software system, the selection criteria in the scoring model should focus on the software's ability to support JIT principles. In Figure 5-3 the three groups of JIT inventory, production, and quality principles presented in this book are listed in a scoring sheet as a part of a scoring model. The particular set of JIT principles can of course be different from the ones based on

FIGURE 5-3 • *Scoring Sheet for Evaluating CIM Software for JIT Integration*

JIT PRINCIPLE EVALUATION CRITERIA	Low support				High support
JIT INVENTORY MANAGEMENT PRINCIPLES					
1. Cut lot sizes and increase frequency of orders	1	2	3	4	5
2. Cut buffer inventory	1	2	3	4	5
3. Cut purchasing costs	1	2	3	4	5
4. Improve material handling	1	2	3	4	5
5. Seek zero inventory	1	2	3	4	5
6. Seek reliable suppliers	1	2	3	4	5
JIT PRODUCTION MANAGEMENT PRINCIPLES					
1. Seek uniform daily production scheduling	1	2	3	4	5
2. Seek production scheduling flexibility	1	2	3	4	5
3. Seek a synchronized pull system	1	2	3	4	5
4. Use automation where practical	1	2	3	4	5
5. Seek a focused factory	1	2	3	4	5
6. Seek improved flexibility in workers	1	2	3	4	5
7. Cut production lot size and setup costs	1	2	3	4	5
8. Allow workers to determine production flow	1	2	3	4	5
9. Improve communication and visual control	1	2	3	4	5
JIT QUALITY MANAGEMENT PRINCIPLES					
1. Maintain process control and make quality everybody's responsibility	1	2	3	4	5
2. Seek a high level of visibility management on quality	1	2	3	4	5
3. Maintain strict product quality control compliance	1	2	3	4	5
4. Give the workers authority to share in the control of product quality	1	2	3	4	5
5. Require self-corrected worker generated defects	1	2	3	4	5
6. Maintain a 100 percent quality inspection of products	1	2	3	4	5
7. Require workers to perform routine maintenance and house cleaning duties	1	2	3	4	5
8. Seek continual quality improvement	1	2	3	4	5
9. Seek a long-term commitment to quality control efforts	1	2	3	4	5

SCORING VALUES ON HOW WELL THE SYSTEM SUPPORTS JIT PRINCIPLE

this book's JIT principles. They should be based on what JIT principles are most important to the company acquiring the software. This scoring sheet in Figure 5-3 also lists a rating score that allows evaluations ranging from 1 to 5. A score of 1 means an evaluator subjectively judges the software as providing low or little support for that particular JIT principle. A score of 5 means the software provides a very high degree of support for the particular JIT principle. Any range of scores can be used in a scoring model (1 to 10, -2 to 2, etc.). Once a scoring system is selected, an evaluator is asked to score each software package for its capacity in supporting the JIT principles. A separate scoring sheet is used for each software system. Once the scores are determined, they are summed, and the software with the highest score is selected. The individual scores by themselves have little meaning and are chiefly of value when used on a comparative basis. Note, under this plan of scoring, each of the JIT principles is given an equal value in the selection process. This assumption may not be as precise or even valid in all situations.

In some situations it might be more precise to weight the importance of the individual JIT principles or groupings of JIT principles (i.e, inventory, production, and quality groupings). To accomplish this weighting with the scoring method, a weight has to be developed. If only one person is making the software decision, he or she can arbitrarily set the weight that person feels best expresses the importance of the JIT principles to the specific organization's success. If multiple evaluators are being used, one simple method to devise a weight is to ask the evaluators who are participating in the JIT software decision to independently rank the importance of principles or groupings of principles to the organization (i.e., a rank of 1 represents a low value of importance). Then, simply sum the ranks and use them as a weight for the principles or groupings of principles. The weighting helps to distribute scoring points in favor of those higher ranked principles. A variety of other advanced techniques can be used to determine these weights, including **multiple regression** and the **analytic hierarchy process** [22].

To illustrate the weighting process with the scoring method, let's look at an example. Suppose a manager in an organization feels JIT inventory principles are of equal importance to both the production and quality principles. The evaluator in this example arbitrarily has attached a weight of 2 to the "inventory" principle scores, while giving a weight of only 1 to the "production" and "quality" principle scores. Using a scoring sheet, like the one in Figure 5-3, the manager scores a JIT software system based on expert judgment on how well the software supports the various JIT principles. These scores are presented in Figure 5-4. Since the three groupings of scores must be individually weighted, they are first individually summed (note "Total Points" in the boxes on the right of Figure 5-4). The summed points representing their scored values are multiplied by their respective weights to obtain the grouped JIT principle scores. The scores are added together to obtain the "Grand Score." In this example the Grand Score of 90 points reflects the double weight given to the importance of JIT inventory principles. It could not be used on a comparative basis with the scores from other JIT software for the final software selection. The software with the highest Grand Score would be the one selected and should be the software that best supports the weighted JIT principles of this example organization.

FIGURE 5-4 • Scoring Sheet for JIT Software Example

| JIT PRINCIPLE EVALUATION CRITERIA | SCORING VALUES ON HOW WELL THE SYSTEM SUPPORTS JIT PRINCIPLE |

JIT INVENTORY MANAGEMENT PRINCIPLES

	Low support →				High support →	
1. Cut lot sizes and increase frequency of orders	1	②	3	4	5	TOTAL POINTS = 20
2. Cut buffer inventory	1	2	③	4	5	
3. Cut purchasing costs	1	2	3	④	5	WEIGHT = × 2
4. Improve material handling	1	②	3	4	5	INV.
5. Seek zero inventory	1	2	3	4	⑤	SCORE = 40
6. Seek reliable suppliers	1	2	3	④	5	

JIT PRODUCTION MANAGEMENT PRINCIPLES

1. Seek uniform daily production scheduling	①	2	3	4	5	TOTAL POINTS = 17
2. Seek production scheduling flexibility	1	2	③	4	5	
3. Seek a synchronized pull system	1	2	3	④	5	WEIGHT = × 1
4. Use automation where practical	①	2	3	4	5	PROD
5. Seek a focused factory	1	②	3	4	5	SCORE = 17
6. Seek improved flexibility in workers	1	2	③	4	5	
7. Cut production lot size and setup costs	①	2	3	4	5	
8. Allow workers to determine production flow	①	2	3	4	5	
9. Improve communication and visual control	①	2	3	4	5	

JIT QUALITY MANAGEMENT PRINCIPLES

1. Maintain process control and make quality everybody's responsibility	1	2	3	④	5	TOTAL POINTS = 33
2. Seek a high level of visibility management on quality	1	2	③	4	5	
3. Maintain strict product quality control compliance	1	2	3	④	5	WEIGHT = × 1
4. Give the workers authority to share in the control of product quality	1	2	3	④	5	QUALITY SCORE = 33
5. Require self-corrected worker generated defects	1	2	③	4	5	
6. Maintain a 100 percent quality inspection of products	1	2	③	4	5	
7. Require workers to perform routine maintenance and house cleaning duties	1	2	3	④	5	
8. Seek continual quality improvement	1	2	3	4	⑤	
9. Seek a long-term commitment to quality control efforts	1	2	③	4	5	

| GRAND SCORE = 40 + 17 + 33 = 90 |

This method has a number of limitations. It is dependent heavily on the evaluator's ability to assess JIT software support, which is not always an easy task. Being a judgmental process, it is open to considerable human bias. This method also requires a valid selection of the JIT principles used as criteria in the model and a valid means of weighting the importance of those criteria. To help overcome these limitations, it is advisable to have a number of individuals perform the scoring to minimize the risk of bias. The **law of large numbers** from statistical theory applies here and would support the notion that the true population evaluation score will stand a better chance of being revealed if a larger number of evaluators are used. The multiple scores can be summed together or averaged to achieve the Grand Score for each software. It may also be helpful to use as large a number of JIT criteria as possible to reveal the total integrative nature of the software being evaluated.

Summary

In this chapter the evolution and basic composition of a CIM system was examined. Depending on the stage of development of the CIM, an organization may have a greater or lesser degree of automation. As presented in this chapter, the basic components of CIMs include computer terminals, robots, AGVs, NC machines, AS/AR systems, electronic sensors, CAD systems, CAE systems, CAM systems, GT cells, and other software. The benefits of combining CIM with JIT were illustrated with an example of how CIM and JIT could be integrated using a software system. An IBM CIM software system that supports JIT principles was used as a basis for the discussion. The chapter ended with a presentation of a methodology that can be used as an aid in selecting JIT software.

To make CIM work, even with JIT principles, takes inventory and scheduling software. One of the inventory and scheduling softwares defined in this chapter as a basic part of CIM is called the "material requirements planning" (MRP) system. The development of MRP systems several decades ago represents a major motivation and a necessary step for the eventual development of CIM systems. The MRP system, though, is based on a large-lot production scheduling operation. This fact appears to many as being a major reason for keeping JIT out of CIM system development. In the next chapter, we will discuss how MRP and JIT can be integrated to improve manufacturing inventory and production scheduling operations.

Important Terminology

Analytic hierarchy process a complex weighting process used to determine mathematical weights for criteria in models that have elements of disproportional importance.

Application System/400™ (AS/400) an IBM brand mainframe computer system.

Artifical intelligence (AI) a non-human form of intelligence. Computers that mimic human decision-making capabilities.

Automated guided vehicle (AGV) a motorized pallet mover that may be centrally controlled or possess independent computer control with limited artificial intelligence.

Automated storage and automated retrieval (AS/AR) a system that performs material handling, order processing, and transportation loading activities.

Backflushing a system of deducting inventory by exploding a product's component parts backwards through its various stages as WIP.

Bar codes a series of alternating bars and spaces printed on items to represent encoded information which can be read by optical scanning equipment.

Computer aided design (CAD) the use of computer software by technicians to create, design and graphically present images of new product ideas.

Computer aided manufacturing (CAM) the use of computer software that can convert CAD designs into computer instructions for automated manufacturing equipment.

Computer aided engineering (CAE) the use of software that takes computer generated information from CAD and allows engineers to conduct simulated tests on design characteristics, such as determining the structural integrity of a design.

Computer integrated manufacturing (CIM) an application of computer hardware and software to achieve an integrated manufacturing system that is controlled in part or wholly by a computer system.

Computer terminal a communication device that is usually hard wired to a computer to provide human users a means to interactive communication and access to data files.

Data base an extensive electronic file of information.

Dedicated automation or machines automated equipment with a limited or single work task capability. The "dedication" of the equipment can also be to a single product or product family. An example is a robot that is only capable of picking up boxes.

Electronic data interchange (EDI) communication between computers using software that permits the exchange of information between differing company communication systems.

Electronic sensors devices used to monitor production equipment or products. They generate data that are conveyed electronically to computer systems where software is used to determine if adjustments are necessary to equipment or products. They can also be used to help robots position WIP in automated equipment like numeric control (NC) machines and in automated storage/automated retrieval (AS/AR) systems.

Flexible manufacturing systems (FMS) a manufacturing layout or system that is designed to permit rapid changeovers to different products and production processes. Usually considered an element of a CIM system.

Foolproof devices automated systems used to monitor and control quality in a production process.

Group technology (GT) a method that seeks to identify the sameness of parts, equipment, or production processes and take advantage of common setup efforts.

GT cells a type of layout facility that permits rapid changes in the sequence of production activities.

Islands of automation an expression used to describe manufacturing facility layouts where multiple or individual pieces of equipment are controlled by a single computer system in a limited domain of a work center or small grouping of work centers.

Kanban a Japanese inventory card system used for controlling manufacturing activities.

Law of large numbers a statistical law that states that larger samples will reveal more information about the true population from which the samples are drawn.

Mainframe computer a term used to describe a large scale computer system, usually the dominant computer system of an organization.

Management information system (MIS) a collection of computer hardware and software used to communicate information within an organization.

Manufacturing Accounting and Production Information Control System/Data Base (MAPICS/DB) an IBM brand of CIM software system.

Manufacturing Automation Protocol™ (MAP) a General Motors system of software standards used to facilitate communication between computer equipment.

Material requirements planning (MRP) a software used to help plan and control the use of dependent demand inventory items and schedule finished product production. (We will be discussing MRP in Chapter 6.)

Microcomputer a term used to describe a small-scale computer, usually networked with other microcomputers to perform a limited number of remote independent applications.

Mini-computer a term used to describe a medium scale computer system, usually used for small business applications or in concert with a mainframe computer.

Mixed model schedule method a scheduling method used where different models of a product family are to be produced by the same production facility or GT cell.

Multiple regression a statistical method used to generate a mathematical expression of a single line. The coefficients of the multiple regression model can be used as mathematical weights reflecting relationships between the variables in the model.

Numerically controlled (NC) machines computer based tools that are programmable to perform a limited number of operating instructions; for example, as an automated drill press.

Optical scanning equipment equipment used to read printed characters or recognize characters by their images.

Optimized production technology (OPT) a scheduling system similar to MRP, but it considers production capacity constraints and schedules finished product production based on optimization procedures.

Process control a management activity of controlling production to achieve a desired output.

Protocol languages computer languages used to convert a variety of automated equipment computer languages into a single language capable of being directed by a central computer system.

Pull system a term used in describing the system by which the use of customer orders pulls inventory through a production system.

Real-time an expression used to describe the lack of lag time between an actual behavior and the report of the behavior being performed from a computer system.

Robot a mechanical device used to pick up, move, or position materials or perform production work tasks.

Scoring model a quantitative evaluation model used to make selection decisions based on criteria judgmentally assessed by experts.

Discussion Questions

1. What is a CIM system? What are the elements that make it up? Explain.

2. How are the objectives of CIM the same as JIT? How are they different?

3. Does a CIM system have to have robots to be considered CIM? Explain.

4. What is the difference between a CAD system and a CAE system? How are they used in a CIM system?

5. What are the benefits of an integrated CIM and JIT system?

6. What are common elements in software systems that support JIT in a CIM environment?

7. What is a scoring model and how can it be used in a JIT operation?

8. Why do we use mathematical weights in scoring models, and where do they come from?

Problems

1. A company must choose a JIT software system from three they have evaluated. The company is using the same scoring system as the example problem in Figure 5-3. JIT system A has received the following scores for its expected ability to support JIT principles: 1, 4, 5, 3, 4, 2, 5, 4, and 5. JIT system B has received the following scores for its expected ability to support JIT principles: 3, 3, 3, 3, 5, 1, 1, 1, and 4. JIT system C has received the following scores for its expected ability to support JIT principles: 1, 1, 1, 1, 5, 5, 5, 5, and 5. Assuming each score is equal in value, which system would you choose? Show your work.

2. A company must choose a JIT software system from five they have evaluated. The company is using the same scoring system as the example problem in Figure 5-3. JIT system A has received the following scores for its expected ability to support JIT principles: 2, 4, 2, 1, 4, 2, 2, 5, 2, 5, 5, and 5. JIT system B has received the following scores for its expected ability to support JIT principles: 3, 1, 2, 1, 2, 4, 5, 5, 1, 5, 1, and 2. JIT system C has received the following scores for its expected ability to support JIT principles: 2, 2, 2, 3, 4, 3, 3, 5, 3, 3, 3, and 3. JIT system D has received the following scores for its expected ability to support JIT principles: 1, 3, 4, 5, 1, 2, 3, 4, 5, 1, 2, and 3. JIT system E has received the following scores for its expected ability to support JIT principles: 5, 5, 5, 1, 1, 1, 1, 4, 4, 4, 3, 3, and 3. Assuming each score is equal in value, which system would you choose? Show your work.

3. A company must choose a JIT software system from three they have evaluated (Software 1, 2, or 3). The company is using the same scoring system as the example problem in Figure 5-3. The three software system scores are presented, respectively, for the following JIT principles: for "Seek production scheduling flexibility:" 1, 4, and 4; for "Cut buffer inventory": 3, 4, and 2; for "Seek a focused factory:" 5, 4, and 5; for "Seek zero inventory": 1, 3, and 2; for "Seek continual quality improvement:" 4, 3, and 2; for "Seek a high level of visibility management on quality": 3, 4, and 5. Assuming each score is equal in value, which system would you choose? Assuming a weight of 3 for JIT quality management principles and a weight of 1 for the other two groupings, which system would you choose? Show your work.

4. A company must choose a JIT software system from six they have evaluated (i.e., Software 1, 2, 3, 4, 5, or 6). The company is using the same scoring system as the example problem in Figure 5-3. The six software system scores are presented, respectively, for the following JIT principles: for "Require self-corrected worker generated defects:" 2, 4, 5, 2, 1, and 6; for "Seek production scheduling flexibility:" 2, 4, 2, 3, 3, and 3; for "Cut buffer inventory:" 3, 2, 1,

3, 3, and 3; for "Seek reliable suppliers:" 3, 4, 2, 4, 3, and 5; for "Seek a focused factory:" 3, 4, 1, 3, 4, and 2; for "Seek zero inventory:" 4, 5, 1, 1, 3, and 2; for "Seek continual quality improvement:" 3, 4, 2, 4, 3, and 2; for "Seek a high level of visibility management on quality:" 3, 4, 2, 3, 4, and 5; for "Use automation where practical:" 3, 4, 2, 1, 1, and 1. Assuming each score is equal in value, which system would you choose? Assuming a weight of 2.5 for JIT production management principles, a weight of 4 for JIT quality management principles and a weight of 1 for JIT inventory management principles, which system would you choose? Show your work.

5. Four managers (Managers 1, 2, 3, and 4) have been asked to choose a JIT software system. Six JIT principles are used as criteria to evaluate the JIT software systems. The managers' scorings of one of the JIT software systems are presented below in the table. The managers were also asked to rank from a low rank of "1" to a high rank of "6" the importance of the JIT principles. These rankings are presented below:

JIT Principles	Scores				Ranks			
	1	2	3	4	1	2	3	4
Seek production scheduling flexibility	2	3	3	2	1	2	1	3
Cut buffer inventory	3	2	4	4	3	1	3	4
Seek reliable suppliers	4	4	4	5	2	3	4	5
Seek a focused factory	3	3	3	4	4	5	5	6
Seek zero inventory	3	4	1	2	5	4	6	2
Seek continual quality improvement	3	4	5	5	6	6	2	1

What is this software system's Grand Score if total points are used? What is the Grand Score if the scores are averaged by principle and then summed? What are the summation of the rankings by JIT principle? What are the average rankings for each JIT principle? Using the summation of rankings as weights for the average scores, what is the software's new Grand Score? Using the average rankings as weights for the average scores, what is the software's new Grand Score? What is the value of all of these Grand Scores? How can they be used to choose a CIM-JIT system?

References

[1] J. B. Dilworth, *Operations Management*, McGraw-Hill, New York, NY, 1992, chapters 5 and 6.

[2] W. J. Stevenson, *Production/Operations Management*, 3rd ed., Irwin, Homewood, IL, 1990, chapter 6.

[3] R. J. Schonberger and E. M. Knod, *Operations Management: Improving Customer Service*, 4th ed., Irwin, Homewood, IL, 1991, chapter 3.

[4] N. Gaither, *Production and Operations Management*, 4th ed., Dryden Press, Chicago, IL., 1990, chapter 5.

[5] L. Mannis, "Extending the Reach Toward CIM," *Manufacturing Systems*, November 1986, pp. 32–36.

[6] D. B. Merrifield, "FMS in USA: The New Industrial Revolution," *Managing Automation*, September 1988, pp. 66–70.

[7] F. Benassi, "Flexible Automation's Star Is on the Rise," *Managing Automation*, September 1988, pp. 28–31.

[8] S. Babbar and A. Rai, "Computer Integrated Flexible Manufacturing: An Implementation Framework," *International Journal of Operations and Production Management*, Vol. 10, No. 1, 1989, pp. 42–50.

[9] A. S. Carrie, et al., "Introducing a Flexible Manufacturing System," *International Journal of Production Research*, Vol. 22, No. 6, 1984, pp. 907–916.

[10] J. I. Finkel, "CIM and JIT—Together They Make the Factory of the Future Work," *Production*, April 1986, pp. 36–40.

[11] J. Romanosky, "Creating the Computer Integrated Manufacturing Factory," *Production and Inventory Review with APICS News*, November 1986, pp. 33–34.

[12] HP Manufacturing Management II, General Information Manual, Hewlett Packard, Cupertino, CA, 1986.

[13] Manufacturing Accounting and Production Information Control System/Data Base, Information Packet, International Business Machines, White Plains, New York, 1989.

[14] G. C. Kim and M. J. Schniederjans, "An Evaluation of Computer-Integrated Just-in-Time Production Systems," *Production and Inventory Management Journal*, Vol. 31, No. 1, 1990, pp. 4–7.

[15] D. W. Rasmus, "Orchestrating Change—Progressing Toward CIM," *Manufacturing Systems*, April 1987, pp. 30–32.

[16] J. Meredith, "New Justification Approaches for CIM," *Journal of Cost Management*, Vol. 1, No. 4, 1988, pp. 15–20.

[17] T. Baer, "Justifying CIM: The Numbers Really Are There," *Managing Automation*, March 1988, pp. 30–35.

[18] A. Young, C. Slem, and D. Levi, "CIM Synergy," *Manufacturing Systems*, March 1987, pp. 16–19.

[19] L. T. Michaels, "A Control Framework for Factory Automation," *Management Accounting*, May 1988, pp. 37–42.

[20] L. C. Giunipero, "Motivating and Monitoring JIT Supplier Performance," *Journal of Purchasing and Materials Management*, Vol. 26, No. 3, 1990, pp. 19–24.

[21] D. M. Cahn, "Performance Measurement: The Key to Success in the 90's," *Production and Inventory Management with APICS News*, Vol. 10, No. 10, 1990, pp. 56–61.

[22] T. H. Willis and C. R. Huston, "Vendor Requirements and Evaluation in a Just-In-Time Environment," *International Journal of Operations and Production Management*, Vol. 10, No. 4, 1990, pp. 41–50.

[23] T. L. Saaty, *The Analytic Hierarchy Process*, McGraw-Hill, New York, NY, 1980.

[24] Arthur Andersen and Company, MAC-PAC/XF Manufacturing and Control System, General Information Manual, Chicago, IL, 1989.

CASE 5-1

Choosing JIT-Ware

Clarkson Industrial Manufacturing (CIM) Limited, of Oakdale, California, owns a highly automated plant in Clever-Ives, Massachusetts (C.I., M.). The plant manufactures up to 47 different industrial plumbing components and operates one of the most state-of-the-art flexible manufacturing systems in the U.S., utilizing MRP, OPT, robots, AS/AR, NC machines, EDI, and bar coding systems. The company's plant currently operates on an intermediate lot size basis with lot runs of as many as 300 units each. The company during the last 30 years has been highly profitable, despite the enormous capital investment in automation. The implementation of their automation has resulted in substantial and continual improvement in product quality during an era where product quality was a key factor to market growth. CIM Limited's "computer integrated manufacturing solution" to product quality and production processing worked well until the early 1990s.

During the late 1980s, the purchasing management of CIM Limited was placed under pressure by vendors to begin a JIT purchasing system of small and more frequent orders. All of the primary CIM Limited vendors switched to JIT inventory methods, and the vendors were losing money on the CIM Limited's large-lot orders. As it turns out, the JIT vendors had to store CIM Limited's order in stock until a sufficient number of smaller JIT lot orders arrived and equaled the desired order size for CIM Limited. These extra storage and handling expenses cost the JIT vendors dearly and could not easily be passed on to CIM Limited. JIT was a strategy vendors all wanted and CIM Limited began to see they would have to live with this fact.

The management of CIM Limited also saw a change in many of their competitors' operations from large-lot to JIT small-lot operations. The quality of the competitors' products also improved and offered CIM Limited more of a challenge in the marketplace than they had experienced in the past. JIT appeared to be a successful strategy for the competitors.

These changes in vendors and competitors worried CIM Limited since they had chosen a different path from the manual methods of JIT, the path of the automated strategy of computer-integrated manufacturing (CIM). Management felt during the 1980s that the CIM system and JIT principles could not be integrated. Management also felt that the considerable investment in CIM systems could not be abandoned simply to try the JIT approach to manufacturing. In the late 1980s new

EXHIBIT 5-1 • *Scoring Sheet for Jack on System X*

JIT PRINCIPLE EVALUATION CRITERIA	SCORING VALUES ON HOW WELL THE SYSTEM SUPPORTS JIT PRINCIPLE				
	Low support ↓				High support ↓

JIT INVENTORY MANAGEMENT PRINCIPLES

	1	2	3	4	5
1. Cut lot sizes and increase frequency of orders	1	②	3	4	5
2. Cut buffer inventory	1	2	③	4	5
3. Cut purchasing costs	1	②	3	4	5
4. Improve material handling	①	2	3	4	5
5. Seek zero inventory	1	2	③	4	5

JIT PRODUCTION MANAGEMENT PRINCIPLES

	1	2	3	4	5
6. Seek uniform daily production scheduling	1	②	3	4	5
7. Seek production scheduling flexibility	1	2	③	4	5
8. Seek a synchronized pull system	1	②	3	4	5
9. Use automation where practical	1	2	③	4	5
10. Allow workers to determine production flow	1	2	3	④	5

JIT QUALITY MANAGEMENT PRINCIPLES

	1	2	3	4	5
11. Maintain process control and make quality everybody's responsibility	1	2	③	4	5
12. Seek a high level of visibility management on quality	1	②	3	4	5
13. Maintain strict product quality control compliance	1	②	3	4	5
14. Give the workers authority to share in the control of product quality	1	2	③	4	5
15. Maintain a 100 percent quality inspection of products	1	2	③	4	5

EXHIBIT 5-2 • *Scoring Sheet for Jack on System Y*

JIT PRINCIPLE EVALUATION CRITERIA	SCORING VALUES ON HOW WELL THE SYSTEM SUPPORTS JIT PRINCIPLE				

JIT INVENTORY MANAGEMENT PRINCIPLES — Low support → High support

1. Cut lot sizes and increase frequency of orders	1	2	3	④	5
2. Cut buffer inventory	1	2	3	④	5
3. Cut purchasing costs	1	②	3	4	5
4. Improve material handling	1	2	③	4	5
5. Seek zero inventory	1	2	3	④	5

JIT PRODUCTION MANAGEMENT PRINCIPLES

6. Seek uniform daily production scheduling	1	2	3	④	5
7. Seek production scheduling flexibility	1	2	③	4	5
8. Seek a synchronized pull system	①	2	3	4	5
9. Use automation where practical	1	2	③	4	5
10. Allow workers to determine production flow	1	2	3	④	5

JIT QUALITY MANAGEMENT PRINCIPLES

11. Maintain process control and make quality everybody's responsibility	1	2	3	④	5
12. Seek a high level of visibility management on quality	1	②	3	4	5
13. Maintain strict product quality control compliance	1	2	3	4	⑤
14. Give the workers authority to share in the control of product quality	1	2	3	4	⑤
15. Maintain a 100 percent quality inspection of products	1	2	③	4	5

EXHIBIT 5-3 • *Scoring Sheet for Iris on System X*

JIT PRINCIPLE EVALUATION CRITERIA	SCORING VALUES ON HOW WELL THE SYSTEM SUPPORTS JIT PRINCIPLE				
	Low support				High support
JIT INVENTORY MANAGEMENT PRINCIPLES					
1. Cut lot sizes and increase frequency of orders	1	2	③	4	5
2. Cut buffer inventory	1	2	③	4	5
3. Cut purchasing costs	1	②	3	4	5
4. Improve material handling	1	2	③	4	5
5. Seek zero inventory	1	2	3	④	5
JIT PRODUCTION MANAGEMENT PRINCIPLES					
6. Seek uniform daily production scheduling	1	2	③	4	5
7. Seek production scheduling flexibility	1	2	③	4	5
8. Seek a synchronized pull system	1	2	③	4	5
9. Use automation where practical	1	2	③	4	5
10. Allow workers to determine production flow	1	2	3	④	5
JIT QUALITY MANAGEMENT PRINCIPLES					
11. Maintain process control and make quality everybody's responsibility	1	2	③	4	5
12. Seek a high level of visibility management on quality	1	②	3	4	5
13. Maintain strict product quality control compliance	1	2	③	4	5
14. Give the workers authority to share in the control of product quality	1	2	③	4	5
15. Maintain a 100 percent quality inspection of products	1	2	3	④	5

EXHIBIT 5-4 • *Scoring Sheet for Iris on System Y*

| JIT PRINCIPLE EVALUATION CRITERIA | SCORING VALUES ON HOW WELL THE SYSTEM SUPPORTS JIT PRINCIPLE |

Low support → High support

JIT INVENTORY MANAGEMENT PRINCIPLES

1. Cut lot sizes and increase frequency of orders — 1 2 3 ④ 5
2. Cut buffer inventory — 1 ② 3 4 5
3. Cut purchasing costs — 1 ② 3 4 5
4. Improve material handling — 1 ② 3 4 5
5. Seek zero inventory — 1 2 3 ④ 5

JIT PRODUCTION MANAGEMENT PRINCIPLES

6. Seek uniform daily production scheduling — 1 2 3 ④ 5
7. Seek production scheduling flexibility — 1 2 ③ 4 5
8. Seek a synchronized pull system — 1 2 ③ 4 5
9. Use automation where practical — 1 2 ③ 4 5
10. Allow workers to determine production flow — 1 2 3 ④ 5

JIT QUALITY MANAGEMENT PRINCIPLES

11. Maintain process control and make quality everybody's responsibility — 1 2 3 ④ 5
12. Seek a high level of visibility management on quality — 1 ② 3 4 5
13. Maintain strict product quality control compliance — 1 2 3 ④ 5
14. Give the workers authority to share in the control of product quality — 1 2 3 ④ 5
15. Maintain a 100 percent quality inspection of products — 1 2 3 4 ⑤

EXHIBIT 5-5 • *Scoring Sheet for Tandy on System X*

JIT PRINCIPLE EVALUATION CRITERIA	SCORING VALUES ON HOW WELL THE SYSTEM SUPPORTS JIT PRINCIPLE				
	Low support ↓				High support ↓

JIT INVENTORY MANAGEMENT PRINCIPLES

	1	2	3	4	5
1. Cut lot sizes and increase frequency of orders	1	2	3	4	(5)
2. Cut buffer inventory	(1)	2	3	4	5
3. Cut purchasing costs	1	(2)	3	4	5
4. Improve material handling	1	2	3	4	(5)
5. Seek zero inventory	(1)	2	3	4	5

JIT PRODUCTION MANAGEMENT PRINCIPLES

	1	2	3	4	5
6. Seek uniform daily production scheduling	(1)	2	3	4	5
7. Seek production scheduling flexibility	1	2	3	4	(5)
8. Seek a synchronized pull system	(1)	2	3	4	5
9. Use automation where practical	1	2	(3)	4	5
10. Allow workers to determine production flow	1	2	3	(4)	5

JIT QUALITY MANAGEMENT PRINCIPLES

	1	2	3	4	5
11. Maintain process control and make quality everybody's responsibility	1	2	3	4	(5)
12. Seek a high level of visibility management on quality	1	(2)	3	4	5
13. Maintain strict product quality control compliance	1	2	3	4	(5)
14. Give the workers authority to share in the control of product quality	(1)	2	3	4	5
15. Maintain a 100 percent quality inspection of products	1	2	3	4	(5)

EXHIBIT 5-6 • *Scoring Sheet for Tandy on System Y*

JIT PRINCIPLE EVALUATION CRITERIA	SCORING VALUES ON HOW WELL THE SYSTEM SUPPORTS JIT PRINCIPLE

	Low support				High support
JIT INVENTORY MANAGEMENT PRINCIPLES					
1. Cut lot sizes and increase frequency of orders	①	2	3	4	5
2. Cut buffer inventory	①	2	3	4	5
3. Cut purchasing costs	1	②	3	4	5
4. Improve material handling	1	2	3	4	⑤
5. Seek zero inventory	1	2	3	④	5
JIT PRODUCTION MANAGEMENT PRINCIPLES					
6. Seek uniform daily production scheduling	①	2	3	4	5
7. Seek production scheduling flexibility	①	2	3	4	5
8. Seek a synchronized pull system	1	2	3	4	⑤
9. Use automation where practical	1	2	③	4	5
10. Allow workers to determine production flow	1	2	3	④	5
JIT QUALITY MANAGEMENT PRINCIPLES					
11. Maintain process control and make quality everybody's responsibility	1	2	3	4	⑤
12. Seek a high level of visibility management on quality	1	②	3	4	5
13. Maintain strict product quality control compliance	1	2	3	4	⑤
14. Give the workers authority to share in the control of product quality	①	2	3	4	5
15. Maintain a 100 percent quality inspection of products	①	2	3	4	5

hope appeared that CIM and JIT systems could be integrated together. A few software developers started marketing a combined CIM-JIT software system that could be used in combination with these two powerful manufacturing strategies. The management of CIM Limited began a software evaluation program to select the best JIT-supporting software system available.

Three CIM Limited production managers (Jack, Iris, and Tandy) were assigned the job of selecting the JIT software. Each of the three was required to attend a seminar on JIT management principles. After the seminar, the three managers were asked to agree on what they felt were the relevant JIT principles for the CIM Limited organization. Fifteen principles were selected for use as criteria in the evaluation and selection process of the JIT software. Currently, only two JIT software systems were considered, System X and System Y. The fifteen JIT principles and the individual rankings of the two software systems by the managers are listed on Exhibits 5-1, 5-2, 5-3, 5-4, 5-5, and 5-6. The three managers were also asked to rank the three groupings of JIT principles (inventory principles, production principles, and quality principles) in the order of their expected importance to the CIM Limited's plant operation. Jack's ranking of the JIT inventory principles is 2, JIT production principles is 1, and JIT quality principles is 3. Iris's rankings of the three are 1, 2, and 3, respectively, and Tandy's rankings are 2, 1, and 3, respectively.

Case Questions

1. What is each software system's Grand Score if total points are used? If you used these values to select the software, what assumptions are you making?

2. What is each software system's Grand Score if the scores are averaged by principle and then the averages summed?

3. What are the summations of the rankings by JIT principle?

4. What are the average rankings for each JIT principle?

5. Using the summation of rankings as weights for the average scores, what are the softwares' new Grand Scores? Based on these values, which software should be selected? If you used these values to select the software, what assumptions are you making?

6. Using the average rankings as weights for the average scores, what are the softwares' new Grand Scores? Based on these values, which software should be selected? If you used these values to select the software what assumptions are you making?

7. Can the CIM Limited's CIM located in C.I., M. be fairly given a JIT by J.I.T. using these methods? Explain.

CASE 5-2

JIT Toward CIM

The Wasson Electronics Company (WEC) of Jackson, North Carolina, is a medium sized electronics manufacturing firm. The company is known for its manufacturing of small electrical appliances, like battery operated radios, phones, ice crushing machines, knife sharpeners, etc. The company has several plants located in the southern portion of the U.S., and has been successfully operating for over 25 years. Sales of the firm's products have been fairly stable over that period and grew according to planned expectations.

Most of the products WEC produces are small and simple in construction, requiring less than 50 component parts. These component parts were purchased from suppliers located throughout the U.S. and Korea. The company's production scheduling system had been computer-based, using an MRP software module. During the 1980s, WEC decided to switch over to a JIT operation. Their stable production rate was ideal for the uniform JIT scheduling operation. Their move was completed by the end of 1987, and all of their plants operated manually, embracing the JIT principles. The only computer systems that were in place in the organization were used by the accounting and marketing departments for order processing and the recording of accounting transactions.

During the late 1980s WEC experienced significant JIT benefits including reduced inventory, improved production processes, and improved product quality. Product quality was so good under JIT principles that finished production inspections were dropped, since they were considered redundant. With the 100 percent inspection of components and WIP, the finished products from these plants simply worked every time.

Most of WEC management was pleased, but the production schedulers who were used to the written production planning and control information expressed concern. They felt the JIT system they had installed worked well because the demand on their products was stable, and the products were small enough to manage with manual methods. They felt any violent shifts in market demand, or the introduction of more complex products, would cause major planning and control problems in production.

Unfortunately, WEC management decided to start production of a very complex electronics system, the "Itsek" systems for U.S. fighter planes. These systems involved tens of thousands of parts and a sequence of production processes that were about a hundred times as complex as the simple products they had been manufacturing. The purchase ordering, delivery timing, and managing lead times of the component parts during the various stages of production would take considerably more effort than the simple manual methods they were using under their present JIT system. Despite the concerns of the production managers, WEC launched into the production of the "Itsek."

Many of the vendors needed to support the "Itsek" product were out of the country and were new to WEC. Problems with currency conversion in paying the vendors got them upset, and some canceled their delivery of ordered components with WEC at critical periods of production. Workers were idle, and they spent more time on housecleaning duties than production time. The plant floors in the JIT operation were so clean they could eat off them. Other vendors made promises but actual delivery dates ended up being different, adding to the confusion of timing and production scheduling. Changes in orders were also having to be made daily and communications with vendors, even within the U.S., took a great deal of time for purchasing staffers, so much so, even managers started helping to write out orders for component parts from vendors.

What units of "Itsek" got finished and delivered to the customers were viewed as poor in quality. The no-check policy used on their other products turned out not to be a good idea for this new product. Customers were considering canceling contracts with WEC.

WEC management was committed to the manufacture of the new product. They realized, though, their version of JIT was not successful when handling a product with long production lead times and requiring a large number of vendors. After discussing the matter with an engineering consultant, WEC management decided that a CIM system, embracing JIT, might be the right kind of system to provide the planning and control necessary for the new product.

Case Questions

1. Describe a complete plan of implementation to take the company from its manual JIT operation to a fully integrated CIM-JIT operation. What are the various stages the company will go through? What are the factors at each stage that should determine a move from one stage to the next? Explain each fully.

2. What specific type of CIM equipment might help with the production problems described in the case? Define each and explain how it would help.

3. What type of CIM equipment might help with their vendor problems? Explain.

CHAPTER SIX

Integrating JIT With Manufacturing Resource Planning (MRP II) Systems

Chapter Outline

Learning Objectives

After completing this chapter, you should be able to:

1. Explain how to integrate a JIT production system into an MRP II production environment.
2. Explain some of the characteristics of a software-supported JIT system.
3. Explain some of the benefits of a combined MRPII/JIT system.
4. Explain the meaning of the terms backflushing, flat bill of materials file, and rate-based master production schedule.

Introduction

In the last chapter we saw how just-in-time methods could be combined with computer integrated manufacturing (CIM) systems. The integration of JIT and CIM permits JIT systems to take advantage of the benefits provided in a timely computer controlled and monitored production environment. While our focus in the last chapter was chiefly on the external CIM system technology, much of the CIM system is controlled internally by **manufacturing resource planning** or MRP II software [1, 2]. Some researchers [3] and software development organizations [4] believe that by integrating JIT with MRP II software, a superior planning and control system will emerge that exploits the benefits of each system while overcoming some of their weaknesses.

The purpose of this chapter is to explain how to integrate JIT with an MRP II system for improved production and inventory planning and control, and to show how an integrated system operates and the informational benefits it can provide to management. To accomplish this objective we will begin this chapter with a brief overview of an MRP system (the precursor of MRP II) and then an overview of an MRP II system. These prerequisite overviews are not designed to be comprehensive, but offer a foundation for understanding the type of planning and control information these systems provide. We will focus on how JIT is integrated into the MRP II environment. Specifically, we will show how the MRP II system supports a JIT operation while providing the computer-based benefits characteristic of the MRP II system.

Overview of Material Requirements Planning

A **material requirements planning (MRP)** system is a computer-based inventory information system used to plan and control raw materials and component parts inventories. It is usually used to plan a future time period (i.e., a forward planning system) of a manufacturing operation, like a month, quarter, or

FIGURE 6-1 • *Overview of the MRP System*

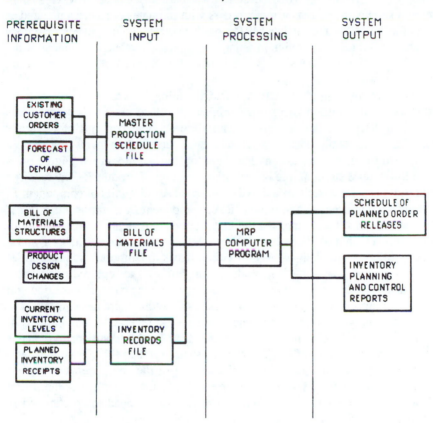

even a year into the future. MRP systems are chiefly used in an **intermittent** (or batch) **manufacturing** operation. Originating in the 1960s, MRP systems were designed to serve the large EOQ lot-size principles of inventory management commonly used at that time.

An overview of the basic components of an MRP system is presented in Figure 6-1. Like all computer-based information systems, MRP systems can be divided into prerequisite information, system input, system processing, and system output.

Prerequisite Information and System Input

The **master production schedule (MPS)** file states the desired finished goods unit production goals in **time buckets** of usually a week. Unlike a JIT inventory pull system that is based on known market demand, an MRP system combines known or existing customer orders with estimates of forecast demand as a basis for the MPS. This makes MRP a **push system,** which pushes inventory through the production operation to satisfy the forecast demand component of the MPS. The MPS file defines for the system what management wants and when management wants it.

The **bill of materials (BOM)** file contains information on how the production of the finished goods is undertaken. A **bill of materials structure** is used (1) to define all the raw materials and component part bill of materials required to complete a product and (2) to describe the multiple levels of assembly or manufacturing necessary to complete a unit of finished product. In Figure 6-2 a typical MRP BOM structure file is presented for a unit of product X. The required component parts for X are two units of component A, four units of component B, two units of component C, and three units of component D. Components Y and Z are given **pseudo bills** to represent the fact that they are an artificial grouping of components (i.e., not a unique component in and of itself). Components Y and Z consist in MRP terminology as **pseudo components** in temporary subassemblies on their way to the next level of product completion. The required component parts to make one unit of Y are two A and four B, while the parts to make one unit of Z are two units of C and three D. We can also see that there are two levels of production activity required to complete product X. At level 2, the components A, B, C, and D are assembled into units of Y and Z. At level 1, components Y and Z are assembled into a unit of product X. We usually refer to the components at a higher level of product completion as the **parent** of the components that go to make it up. For example, X is the parent for components Y and Z, Y is the parent of A and B, etc. One or more of these components may need to be changed in some future time bucket to reflect product design changes.

The inventory records file of the MRP system defines current levels of finished goods, raw materials, and component parts inventory at the beginning of some planning period. During the planning period, the organization may receive units of raw materials, component parts, subassemblies, and even finished goods inventory from suppliers, vendors, and subcontractors. These planned inventory receipts and delivery lead times are included in the inventory records file so their addition can be appropriately considered in the time bucket of their arrival.

System Processing and System Output

Once the MRP system knows what it is expected to produce (via the MPS file), how it should produce it (via the BOM file), and with what it has to produce it (via the inventory records file), the system arithmetically combines the information to determine when the production should take place in the future planning period. To accomplish this a process called **requirements explosion** is

FIGURE 6-2 • A Multi-level MRP Manufacturing Bill of Materials Structure File for Product X

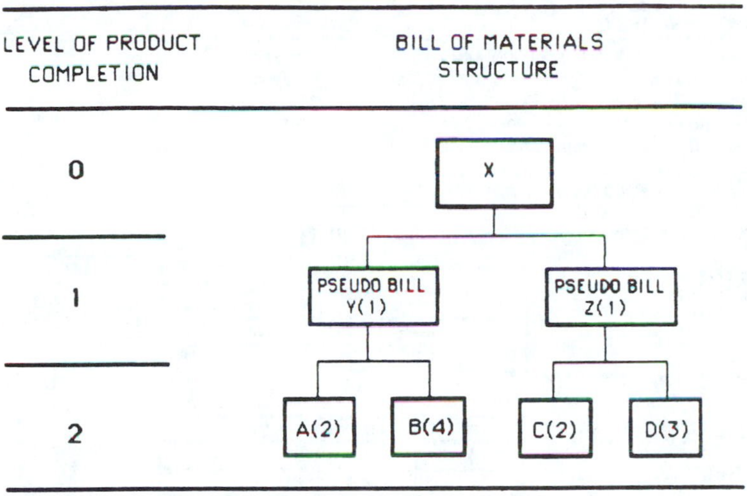

LEVEL OF PRODUCT COMPLETION	BILL OF MATERIALS STRUCTURE

conducted. The program starts with the finished goods demand from the MPS and "explodes" the demand requirements backward in time to schedule the desired production of the finished goods from raw materials and component parts with "time-phased" adjustments for lead time requirements. Let's look at an example of requirements explosion. Suppose we want to produce 100 units of product X in accordance with its inventory requirements expressed in the BOM file in Figure 6-2. To explode the component part requirements backward through time, we need to establish some lead time requirements for the assembly and vendor ordering activities that make up product X. These lead time requirements are presented in Figure 6-3(a) and the requirements explosion is presented in Figure 6-3(b).

As a result of the MRP computer program requirement explosions, management is provided with inventory management information for planning future production. The MRP system provides a schedule of "planned order releases" of inventory to allow managers to see how well the desired MPS file objectives for finished goods will be achieved in the future planning period. Many organizations operate their MRP systems so that workers log-on and log-off job orders directly on the computer.

During the use of the MRP plan, management monitors the actual weekly operations against the MRP plan. To assist in helping management to comply with the MRP plan, the MRP system also provides detailed inventory planning and control reports on inventory status, exceptions to stated goals (e.g., missing a lead time), and capacity loads in departments or at work centers.

FIGURE 6-3   •   *Example of MRP Requirements Explosion for Product* X

(a) Lead time requirements for inventory

PRODUCTION ACTIVITY	REQUIRED LEAD TIME
Assemble 100 units of X from 100 Y and 100 Z	2 weeks
Assemble 100 units of Y from 200 A and 400 B	2 weeks
Assemble 100 units of Z from 200 C and 300 D	1 week
Receive 200 units of A from vendor	1 week
Receive 400 units of B from vendor	1 week
Receive 200 units of C from vendor	1 week
Receive 300 units of D from vendor	1 week

(b) Chart of lead time adjusted materials requirements for product X

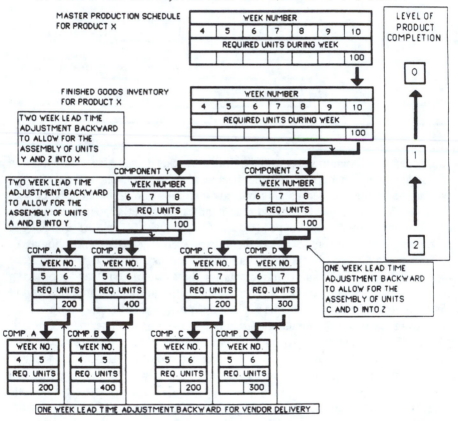

FIGURE 6-4 • *An Overview of an MRP II System*

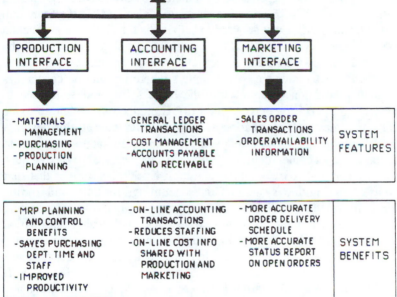

Overview of an MRP II System

As MRP systems became more popular in the 1970s and 80s, the need to expand the benefits of using such a system into other areas became apparent to many software developers. MRP systems were expanded to share information with purchasing, accounting, and even marketing departments. This new enlarged role for MRP systems necessitated a new name: **Manufacturing Resource Planning** or **MRP II.** Most MRP II systems are a collection of computer programs or system modules that permit the sharing of information within and between departments in an organization. An overview of a typical MRP II system's features and benefits is presented in Figure 6-4.

Production Interface

In addition to an MRP module that provides all the inventory planning and control features of an MRP system, most MRP II systems also have purchasing and production planning modules. The purchasing module permits the BOMs ordered by the MRP module to be automatically printed and sent to vendors and suppliers without the need for purchasing agents. While most MRP II systems permit purchasing managers to monitor and control such automatic purchasing systems, much of the tedious paper work (and therefore staff) is saved by sharing information between the MRP and purchasing modules. MRP II production modules are software systems that are used to help in the scheduling of production activities. Some production modules allow managers to simulate changes in the MPS to see the impact on production scheduling while other modules are used to identify production departments or work centers where production capacity is limiting overall production output. While the basic MRP module is used to support these types of decision making, the MRP II production modules provide considerably more detailed information at each **product completion** level. A typical MRP II production module might include software support systems to establish a job order schedule, route the job through the necessary work centers, track the order on an on-line basis, and even establish priorities for jobs that start forming bottlenecks in the system. This scheduling, routing, and timing information is provided in on-line or hard copy reports that help management complete their system planning and control objectives. Such timely information helps management to locate capacity limitations in the production departments and make decisions to prevent them from idling capacity in other departments, thus improving a system's productivity. Most important, in the MRP II environment the relevant information for these transactions of inventory and production activity are shared with others in the organization that must use and process the information.

Accounting Interface

As inventory is acquired from vendors and suppliers, accounts payable (A/P) must be adjusted to reflect the debt. The MRP II interface of the purchasing department allows vendor and supplier information (such as address, units ordered, cost, etc.) to be automatically passed to A/P to be recorded without the need for accounting staff support. As inventory passes through its various **work-in-process (WIP)** stages, the cost of the inventory used, not used, and wasted must be updated to reflect current manufacturing costs. The MRP II interface with the production modules allows this cost information to pass directly to the accounting department computer interface so accurate and up-to-date cost information can be provided to production managers for planning and control purposes. By having up-to-date cost information on products, managers can make timely product cancellation decisions on products whose current costs exceed their marketability. As finished goods inventory is sent out to customers, accounts receivable (A/R) must be adjusted to reflect the future receipt of the sales funds. The MRP II interface not only records the future A/R from the customer, but also provides this information to the marketing department so sales personnel are up-to-date and can better control the amount of credit issued to their customers.

Marketing Interface

While the basic MRP system provides fairly accurate information on the delivery of customer orders, that information is usually available to marketing personnel only when passed on by a production manager. In the MRP II environment, the marketing personnel can access that information directly with their own computer terminals or ports, saving time that may help make a sale. Marketing personnel can also enter possible customer orders and receive estimated order release information without the assistance of any production personnel. Being able to provide customers with a fairly accurate estimate of an order delivery immediately and without needing to go through production management greatly helps in relieving customers' doubts and improves the chances of making sales. The improved accuracy in delivery times saves companies the costs (i.e., wasted materials, labor, etc.) of jobs that are later canceled by customers who were given impossible delivery dates. Marketing staff can also check the status of WIP and job order priorities for open order jobs in production queues. Being able to provide this information to customers on an on-line basis represents a significant competitive advantage over other less integrated or non-computerized systems.

In summary, the MRP II system is an enhanced MRP system that integrates various departmental activities to permit a greater sharing of planning and control information. The result of this shared information is improved efficiency by making timely and more knowledgeable decisions that save the material and human resources of the organization.

Integrating JIT and MRP II

Some researchers feel that JIT is chiefly limited in application to a **continuous flow** or **repetitive** type of manufacturing operation while MRP and MRP II are more appropriate systems for an **intermittent** type of production operation [5]. It is believed that the continuous flow type of production laid out in a sequence of production activities like an assembly line takes maximum advantage of the **synchronized production** characteristics (we discussed these in Chapter 3) of JIT inventory usage and production scheduling. It is also believed that the intermittent type of production where large-lot batches or production runs are laid out in departmentalized functional areas takes maximum advantage of the planning and control characteristics of MRP and MRP II systems. Other researchers feel JIT has some application in any type of production operation [6] and the best system would be a mixture or joint MRP II and JIT system capable of handling all types of production operations [3, 7, 8].

To integrate JIT in an MRP II environment would require the use of a software system that would support JIT production while still offering users the on-line benefits of the MRP II system. One of the first software developers to offer a mixed-mode system that combines MRP II and JIT was Hewlett Packard (HP) with the development in the mid 1980s of their MRP II system called HP Manufacturing Management II [4]. This MRP II system allowed users to add a JIT module to eight other modules (materials management, sales order management, production management, production cost management, purchasing, accounts payable, accounts receivable, and general ledger modules) that comprise the MRP II system. The HP JIT module permits manufacturers to take advantage of traditional JIT concepts, including worker involvement, demand pull production, and cooperative vendor relationships, while providing MRP II on-line planning and control information on production activities. There are other MRP II/JIT software systems than the one developed by HP, and they share a number of common characteristics.

Characteristics of MRP II/JIT Systems

The five MRP II/JIT system characteristics presented here are characteristics required to allow the JIT software system to support JIT functions. These five system characteristics include:

1. Flat BOM files
2. Stock areas
3. Deduct points and lists
4. Rate-based master production scheduling
5. Interface with accounting transactions

To be able to integrate MRP II and JIT systems requires the user to understand how they support JIT production operations.

1. Flat BOM Files Unlike the multi-level BOM file in the MRP II system JIT software requires a **flat BOM file** possessing only a single level of product completion. An example of a single-level BOM file for product X is presented in Figure 6-5. Comparing the MRP BOM file from Figure 6-2 with that of Figure 6-5, we can see that each requires the same number of components for a finished unit of *X*, except for the pseudo bills (components) representing intermediary production steps. The flat BOM is designed to be simple and take advantage of the continuous flow, unitary production characteristics of JIT.

FIGURE 6-5 • *A Single-level JIT Manufacturing Bill of Materials File for Product X*

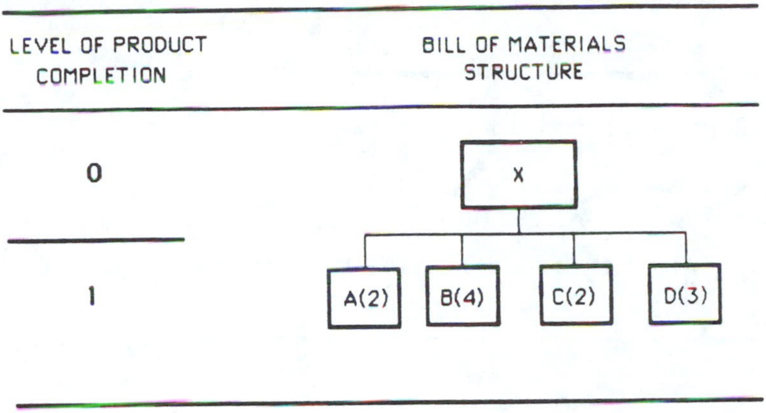

LEVEL OF PRODUCT COMPLETION	BILL OF MATERIALS STRUCTURE
0	X
1	A(2) B(4) C(2) D(3)

2. Stock Areas Raw materials or component parts inventory must be brought into the manufacturing area and temporarily stored for use. In the MRP system, large lot sizes necessitate large storage areas. In a JIT operation unitary or small lot sizes are brought to **stock areas** within the manufacturing facility. A stock area can be in a work center, stock room, work cell, or just a place along a production line. A stock area is an area where inventory will be used in the production process. An example of a series of stock areas needed to manufacture product X is presented in Figure 6-6. We can also see that the total number of components of A, B, C, and D necessary to complete one unit of product X, as stated in Figure 6-5, are the same as required in Figure 6-6. The unitary nature of JIT production requires the assembly of each unit of product X to flow in a step-wise nature from stock area 1 to 4. The output subassembly from each prior stock area becomes a unit of input for the next stock area. The stock area inventory output is basically the same as a pseudo component used in MRP systems. Most MRP II/JIT software systems are designed to share the common MRP II/JIT data base when determining inventory requirements. This allows the joint system to permit typical MRP inventory transactions and the integrated accounting system to keep track of them.

Another advantage of the use of the stock areas is that they can be used as destination points for the implementation and support of kanban card systems. As we saw in Part 1 of this book, kanban card systems are beneficial inventory support systems that help JIT systems achieve their unitary production goals. Indeed, setting up a kanban post at each stock area would allow the MRP II system in this example to convert completely to a traditional JIT system not in need of computer support. As previously stated, the joint MRP II/JIT system can be used as an implementation strategy to move to the traditional JIT operation.

FIGURE 6-6 • *Illustration of Stock Area Design for Product* X

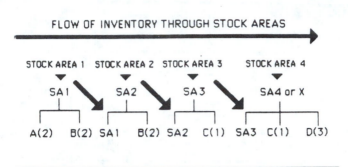

3. Deduct Points and Lists   A **deduct point** is a point in the production process where raw materials and/or component parts are removed, or deducted, from the record of existing inventory. A deduct point is usually a stock area. As an inventory item passes through a stock area, the amount of raw materials or component parts that make up the parent are deducted from inventory. This means that for the JIT module, inventory accounting is usually done after the finished product is completed. The process of deducting the component inventory after it has been used in the completion of a finished product is called **backflushing** or a "post-deduct." In Figure 6-7 we can see an example of this backflushing for product *X*. If we remove a unit of product *X* from stock area 4, the components that make up one unit of product *X* are deducted. The number of items used to make one unit of the parent at each stock area are found on that stock area's deduct list. A deduct list is basically a BOM for each stock area. The deduction of product *X* causes the system to record a deduction of one unit of *C*, three *D*s, and one unit of output from stock area 3 or SA3. The deduction of SA3 necessitates a deduction from inventory of one unit of *C*, and one unit of output from stock area 2 or SA2. This process of backflushing of inventory continues until all of the necessary components for one unit of product *X* have been removed or decremented from existing inventory records. This is quite different from an MRP system which would require work orders and their inventory stock requisitions to trigger inventory transaction accounting. Most important, though, is the fact that deduct points allow for unitary reductions in inventory. We don't have to be locked into the MRP large lot-size production activity in planning the future because the JIT unitary deductions in inventory can be changed on a current daily basis if shifts in demand warrant the changes.

While kanban systems can operate without the expense of a computer system, greater control on the status of inventory can be implemented in the JIT system by having more deduct points along the production line and employing computer-based inventory recording equipment like bar code scanners at each stock area. As the inventory passes through each stock area, workers use the scanners to record the conversion of inventory. As the number of stock areas or deduct points increases and inventory is scanned and recorded at each deduct point, management will be able to more timely keep track of the reduction of WIP component parts inventory.

4. Rate-Based Master Production Scheduling   One of the most important JIT features is its flexibility in dealing with changes in actual demand on a timely basis. Specifically, the scheduling of production in a JIT system is related to the demand pull. Most users of MRP systems agree that a minimum production scheduling time bucket is a week, and that a future planning horizon is usually a month, a quarter, or even a year. MRP II/JIT software is designed to support the daily fluctuations in production that would be used to determine the rate of production in subsequent days.

A **rate-based master production scheduling** system fluctuates to meet the demand pull of a JIT system. As in an MRP system, the rate-based master production schedule is used to set monthly production goals or quotas for the JIT operation to meet. As stated in Chapter 3, production quotas are used to motivate production effort in JIT operations. Also as in an MRP system, the monthly product demand can be based on both known customer orders and forecast demand. Of course we know that the less demand is based on forecasts and the more it is based on actual known customer orders, the more the operation will benefit from the reduction of inventory that is derived from the JIT approach.

FIGURE 6-7 • *Illustration of Multiple Deduct Points for Product* X

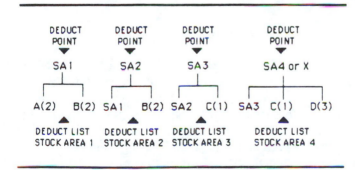

TABLE 6-1 • *Rate-Based Master Production Schedule for Product* X

	MONTH		
PRODUCTION/INVENTORY ITEM	*JAN/1991*	*FEB/1991*	*MAR/1991*
Number of working days	22	20	22
Production rate (units/day)	10	9 (decrease)	11(increase)
Backlog of units (for month)[1]	40	20	0
Unit forecast (for month)	200	150	250
Total units (for month)[2]	40	170	50
Production (units/month)[3]	220	180	24
Beginning inventory (for month)[4]	22	2	12
Available inventory (for month)[5]	242	82	254
Ending inventory (for month)[6]	2	12	4

1. Backlog of units = last month's total units–last month's available inventory (can also be assigned from any number of prior month's capacity shortage as presented in this example where the 40 and 20 units are from production shortages for months before Jan. 1991)
2. Total units = backlog + unit forecast
3. Production = number of working days x product rate
4. Beginning inventory = last month's production–last month's total units (or last month's ending inventory)
5. Available inventory = production + beginning inventory
6. Ending inventory = this month's production–this month's total units

Unlike the MRP system, daily shifts in the production schedule are allowed and even encouraged in response to daily shifts in actual product demand. In Table 6-1 a rate-based master production schedule for product X is presented. In each month a total demand in units is established as a goal. Total production in units is then established as a response to meet the shifts in total demand. The shifts in the production rate are necessary each month to produce enough units to meet the required total units for the month. The shifts in the production rate are accomplished by adjusting capacity to meet demand requirements. By providing future production rates necessary to meet expected future demand, the JIT module assists management in rough-cut capacity planning. As with a JIT system, the production rate can be shifted on a daily basis if the difference between the MPS and actual demand warrants the change. These daily production rate shifts are implemented at the shop floor level by computer-based reports (either hard copy

TABLE 6-2 • *A Summary of System Characteristics*

Characteristic	*MRP II System*	*JIT System*	*MRP II/JIT System*
BOM File	Multi-level	Single-level	Either or both
Stock Areas	In departments or work centers	In work centers	Either or both
Master Production Scheduling	Schedule fixed for future planning period (a forward planning system)	Rate-based schedule flexible to meet changes in demand (a current period planning system)	Either or both
	Minimum planning period one week	Minimum planning period one day	Either or both
	Based on known demand and forecasts	Based on known demand	Either or both
	A push inventory usage system	A pull inventory usage system	Either or both
Type of Operation	Chiefly intermittent	Chiefly continuous flow	Either or both
Type of Management Control Reports	On-line and detailed	Physical observation with some computer application for JIT card systems and bar coding	On-line and detailed but capable of supporting JIT card systems and bar coding Inventory
Inventory Accounting	On-line and detailed	Physical card system with some computer application	On-line and detailed, capable of supporting card data entry systems
	Based on stock requisitions before production starts	Based on a physical audit of kanban cards as production occurs	On-line, based on a pooled data base that uses stock requisitions for MRP production and backflushes stock for JIT production

or on-line through a computer) to work centers. Changes are also made to the rate-based MPS to take the current changes into account in future months. In this way the computer-based MRP II system supports the flexibility found in JIT, but is implemented more quickly as the on-line system found in MRP II.

Since the production rate is determined on a daily basis, JIT materials requirement planning does not compute planned order releases for inventory items in advance. Instead, the JIT module determines on a daily basis the availability of just those raw materials and component parts required to support the rate-based MPS. It is essential that JIT vendors and suppliers be used to support this system in order to reap the reduced inventory benefits of the JIT approach.

5. Interface with Accounting Transactions Non-computer-supported JIT operations in the U.S. are required to perform considerably more auditing and accounting transactions for governmental purposes than their Japanese counterparts. While some of these accounting transactions are accomplished using computer applied technology, an integrated system that shares a common accounting data base is far more efficient than individual computer applications that must be down-loaded or up-loaded at a later date. Most MRP II/JIT software takes full advantage of the other MRP II accounting modules that record inventory transactions for general ledger and costing purposes. Indeed, JIT software supports the same materials cost reports and exception reports that are provided in the MRP II system, while supporting computer-based auditing functions like "cycle-counting."

A summary of differences between MRP II, JIT, and joint MRP II/JIT systems is presented in Table 6-2. As we can see, a joint MRP II/JIT system can support either system separately and has a number of advantages depending on the type of operation and production situation in which it is applied.

A Strategy for Implementing JIT in an MRP II Environment

Some manufacturing operations, such as those in fabrication and prototype development, will always have intermittencies that are best handled by batch-type MRP and MRP II systems. Yet competition now and in the future continues to force organizations to adopt systems like JIT that permit greater flexibility [9]. The Hewlett Packard HP JIT module of their MRP II system was specifically designed to assist users who are interested in converting some of their operations to a JIT format. In this sense, the HP JIT software was designed as an implementation strategy for integrating JIT into an MRP II operation. But just acquiring JIT software will not ensure the successful adoption or integration of JIT into an MRP II environment.

For a successful integration of JIT into an MRP II environment management must develop a strategy that embodies all of the JIT principles discussed in Chapters 1 through 4. The conversion to a software-supported JIT operation from the typical intermittent operation of an MRP II environment requires several changes, including increased flexibility in purchasing policies with vendors, increased flexibility with workers, a synchronized facility layout and an alteration in ordering policies with customers. Let's look at how each of these changes can be implemented by way of a comparative example. For purposes of the illustration, let's say we are going to convert an all-MRP II operation into a computer-supported JIT operation.

In Figure 6-8 a comparison of inventory flow layouts under MRP II and JIT systems is presented for product X. In the MRP II layout in Figure 6-8(a) a lot of 100 units of product X is produced every ten days. Using the Figure 6-2 BOM file (disregard the lead times in Figure 6-3 for this example) the company receives the necessary component parts for 100 units of product X from its vendors every ten days. The assembly work in the typical MRP II intermittent operation is departmentalized and the units of inventory are moved on a lot-size or batch usage basis. The necessary component parts A and B move from the receiving department to assembly department 1 to permit the assembly of the pseudo component Y, and likewise, the necessary component parts C and D move from the receiving department to assembly department 2 to permit the assembly of the other pseudo component Z. From assembly departments 1 and 2 the 100 component parts each of Y and Z are sent to assembly department 3 for final assembly into 100 units of product X. From the assembly department 3 the 100 units of X are moved to finished goods inventory. When a customer order requires them to be moved to the shipping department, a lot-size order of 100 units is shipped to a customer at the end of the ten-day period.

Throughout the MRP II layout in Figure 6-8(a) the inventory components have been moved in a 10-day batch requiring temporary storage space in each department. To move the lot-size batches of inventory in this layout would require relatively more material handling staff and equipment than a comparable JIT operation. Also the typical MRP II system requires a finished goods inventory to permit the finished product to wait until actual product demand catches up with the forecast "pushed" inventory production. One basic element of implementing JIT is to move toward a continuous flow production operation and away from a batch type of operation. To implement this change toward a JIT operation involves the use of four tactics:

1. Increased flexibility in purchasing policies with vendors
2. Increased flexibility with workers
3. A synchronized layout
4. An alteration in ordering policies with customers

Collectively these four tactics make up a strategy for integrating JIT and MRP II. Let's look at each of the four tactics.

1. Increased Flexibility in Purchasing Policies with Vendors As we discussed in Chapter 2, a move toward JIT necessitates the synchronization of the

FIGURE 6-8 • Comparison of Inventory Flow Under MRP II Intermittent and JIT Synchronized Production Operations

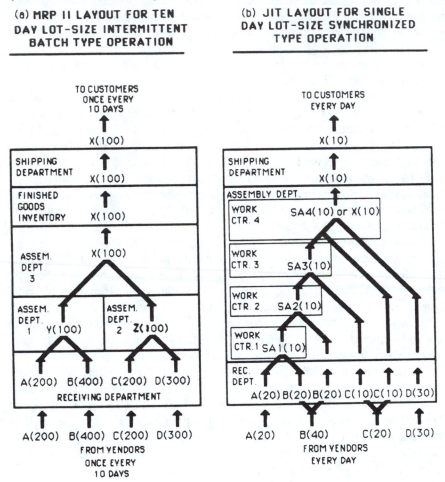

(a) MRP II LAYOUT FOR TEN DAY LOT-SIZE INTERMITTENT BATCH TYPE OPERATION

(b) JIT LAYOUT FOR SINGLE DAY LOT-SIZE SYNCHRONIZED TYPE OPERATION

flow inventory from vendors to match daily (or even hourly) production. An initial strategy of change might be to shift from a 10-day lot-size order policy in Figure 6-8(a) to a 1-day lot-size order policy in Figure 6-8(b). As we can see in Figure 6-8(b), the number of units from the vendors arrive daily at a daily usage rate of 10 units (10 percent of the MRP II 100 unit lot-size). This reduces the need for about 90 percent of the inventory staging area of product X in the receiving department. The future lot-size order policy should be negotiated with vendors to allow daily variable lot-size order changes. By allowing variable or floating daily lot sizes the organization can shift inventory usage to meet the actual pull of inventory by demand in the JIT system. By building the flexibility in inventory ordering policies the reduced inventory benefits of JIT (discussed in Chapter 2) can be accrued in this type of operation.

2. Increased Flexibility with Workers As we discussed in Chapter 3, a move toward JIT necessitates greater flexibility in worker job assignments. As demand shifts on a monthly, weekly, or daily basis, management must have the flexibility to increase or decrease its production resources to synchronize production rates with capacity requirements. This permits the software-supported JIT operation to accrue the reduced cost benefits of eliminating idle or wasted labor resources that would otherwise go into units of production that are not in demand. One of the methods that can be used by management to facilitate the worker crew size decision during shifts in demand is a **cycle time/crew size matrix** [10]. The cycle time/crew size matrix is a crew size determination schedule based on the cycle time required for production. To develop the matrix requires a determination of the cycle time and labor content for each of the products a facility produces. In the example in Table 6-3 the **cycle time** for completing one unit of product X, based on completing 100 units of product X in 10 days, with an assumed 8 hours per day is 2,880 seconds, or:

Cycle time
for product X =   Total seconds of time for production in 10 days
                              Total unit production in 10 days

               =   60 sec./min. x 60 min./hr. x 8 hrs./day x 10 days
                                    100 units

               =   2,880 seconds per successive completion

The **labor content** is the investment of labor time per unit of production for a constant or given manufacturing process. The labor content will remain constant as long as the manufacturing process remains constant. Allowing an individual worker for each of the work centers in Figure 6-6, we can restructure the department layout into work centers with stock areas as presented in Figure 6-8(b). The figure shows there are four required workers, and each will occupy one of the four work centers. Given a four-worker manufacturing process, the labor content for each of the 100 units that they can produce in a 10-day period is 11,520 seconds, or:

Labor content
for product X   =   Number of workers x Total seconds
                                  of labor time for production in 10 days
                                  Total unit production in 10 days

                         =   4 workers x 60 sec./min. x
                                  60 min./hr. x 8 hrs./day x 10 days
                                  100 units

                         =   11,520 seconds of labor per unit

The ratio of the labor content and cycle time statistics can be used to determine crew size. Labor content remains constant for a given manufacturing process, but cycle time changes in a rate-based production system to meet changes in demand. The ratio of a given labor content to a demand "pulled" cycle time determines the crew size necessary to produce the demand requirements. In Table 6-3, the cycle time/crew size matrix for product X for the JIT operations is presented. The matrix provides an easy-to-use guide that facilitates crew size decisions.

Workers also need to be cross-trained and contractually flexible to handle a variety of jobs to accommodate the changes required in a rate-based production system. The nature of the intermittent operation necessitates a variety of differing production job runs and their setup time costs. To handle the occasional bottleneck caused by unbalanced work assignments, management must have the flexibility to reassign and alter jobs to improve inventory flow. The movement from batch production to continuous flow production will mean the elimination of departmentalization of job skills. Workers in a software-supported JIT operation will have to be cross-trained in the multiple skills that were previously defined by

TABLE 6-3 • Cycle Time/Crew Size Matrix for Product X

Product X Unit Demand	Labor Content +	Cycle Time	=	Crew Size
250	11,520	1,152		10
225	11,520	1,280		9
200	11,520	1,440		8
150	11,520	1,920		6
100	11,520	2,880		4

their departmental affiliations. In Figure 6-8(b) we can see the work activities of the three assembly departments for the MRP II operation are restructured according to the four stock areas design presented in Figure 6-6. Cross-training also permits management to quickly reassign workers to bottlenecked work centers without losing time to retraining. The cross-trained worker's increased job knowledge may enhance the capability to make cost-reducing job run setup recommendations that are characteristically observed as benefits in traditional JIT operations.

3. A Synchronized Facility Layout   The physical layout of the MRP II operation must be changed to support the JIT type of continuous flow production system, thus necessitating a "uniform daily production" rate in units. This might require the dividing of departmental work activities among a series of work centers. Like an assembly line, the sequence of work activities is performed on each unit, one at a time. In Figure 6-8(b) we can see the work activities of the three assembly departments for the MRP II operation are restructured according to the four stock areas design presented in Figure 6-6.

This change in layout is necessary to support the rate-based production schedule required in JIT continuous flow production system. In an MRP II batch mode a component part inventory allocation for the entire 100-unit order of product X is made ahead of actual demand in the example in Figure 6-8(a). So if the demand changes during the ten-day production period, the operation is not flexible enough to make a timely adjustment to production. Indeed, if demand decreases below the planned 100-unit lot size, management would have to choose to continue the current 100-unit lot size production run and store the unneeded units in finished goods inventory or suspend production and incur WIP inventory costs on the unneeded subassemblies until the MRP II system could regenerate a new plan. In the case of an increase in demand beyond the 100-unit lot size, management would have to choose to backorder the additional surplus demand until the next production run where again the MRP II system would generate a new plan with increased lot sizes or incur the costs of letting a competitor subcontract the excess in demand. In the software-supported JIT operations in Figure 6-8(b), the layout permits unit production to be changed on a daily basis. If a decrease of 16 units in the 100-unit order for the ten-day demand period is dictated on day four, then production in each of the remaining four days of the ten-day production run can be reduced from ten to six units per day. The reductions in staff and inventory necessary to support the reduced production level allows the system to avoid the costly waste of needless inventory and labor usage. Similarly, the JIT system could accommodate increases beyond the 100-unit order demand level on a daily basis without incurring the costs of the less flexible MRP II system.

The change in layout is also needed to facilitate inventory accounting and control. In Figure 6-8(b) the small lots of component parts inventory are placed in each storage area daily. As the inventory items are used to assemble a unit of finished product, they pass through each work center. Consistent with Figure 6-7, each work center in Figure 6-8(b) is a deduct point where inventory for each parent is deducted from its respective storage area in accordance with each deduct list. The HP JIT module is capable of deducting the component inventory usage either when the finished item is completed or when inventory is used at each stock area to complete a parent. The HP JIT module provides daily computer reports updating inventory status which can be easily matched up by management with their location in the physical facility for auditing and control purposes. The HP JIT module provides the same accounting (and marketing) reports as those obtained for planning and control in MRP II.

4. An Alteration in Ordering Policies with Customers   Just as we must develop a JIT ordering policy with the vendors who supply the component parts inventory, we must establish an ordering policy with our customers to avoid a finished goods inventory once finished products are released from the system. Strategies for establishing ordering policies with customers are discussed in Chapter 2, including special discounts as incentives to accept the smaller and more frequent JIT orders. As we can see in Figure 6-8b, the customers in this example would be required to accept a daily order of ten units of product X instead of the 100-unit order each ten days.

In summary, the conversion of an MRP II system to a computer-supported JIT system should generate the same amount of production in a ten-day period, but with a reduction of component parts and finished goods inventory of about 90 percent. While this type of inventory reduction benefit is common to the traditional non-computer-supported JIT system, the computer supported JIT system is the only system that provides the detailed inventory status reports, load reports, and exception reports that typify the forward planning MRP II system. The inventory status reports are generated on a daily basis (not weekly like an MRP II system) and let management know what the current status of inventory (of raw materials, component parts, subassemblies, and finished goods) is in the system, permitting them to better monitor and manage exceptions. The computer-supported JIT system is also more flexible than the MRP II system in allowing for daily shifts in production to meet shifts in demand. Moreover, the forward projection of the production schedule under the rate-based computer-supported JIT system permits management an opportunity to anticipate changes and respond to major shifts in demand (where forecasting is used in the MPS), as opposed to the wait-and-see approach of the traditional JIT system. The joint MRP II/JIT system takes advantage of the flexibility of JIT and incorporates it with the forward planning and on-line timely reporting advantages of an MRP II system.

Summary

This chapter has presented a review of the basics of MRP and MRP II systems, and explained some of the characteristics and benefits of a combined MRP II/JIT production system. The intent of the chapter was to show how the inflexibility that exists in a typical MRP II system can be made more flexible by the incorporation of a computer-supported JIT system. An example was provided to illustrate some of the conversion strategies and efforts that are necessary to implement the computer-supported JIT system in an MRP II environment.

The content of this chapter has focused chiefly on the Hewlett Packard MRP II/JIT system. In fairness to other computer system developers, like IBM, it should be mentioned that most major software developers have MRP II systems that can be used to support JIT. Unlike the HP MRP II/JIT system, many of the other developers do not have separate JIT modules. Instead, software like IBM's Communications Oriented Production Information Control System (COPICS) provides related software modules that can be modified to support some of the JIT rate-based scheduling activities needed for a JIT system [11]. Adoption of such joint MRP II/JIT systems should be based on how well information is shared and used to support the activities between the MRP II and JIT systems. (The model presented in the "JIT Software Selection Methodology" section in Chapter 5 can be applied to help select an MRP II software system that best supports JIT principles.)

Important Terminology

Backflushing a system of deducting inventory by exploding the bill of materials backward from a unit of finished goods. Also called "post-deduct," this system reduces the component inventories only after the parent units are finished.

Bar code a system used to identify an inventory item by using electronic or optical scanning equipment. A label placed on inventory items that is made up of a series of alternating bars and spaces that is encoded by computers. It is used to facilitate the timely and accurate entry of inventory transactions in computer-based systems.

Bill of materials (BOM) a listing of subassemblies, pseudo assemblies, pseudo components, component parts, and raw materials that go into a parent assembly or parent component showing the quantity of each required to complete its respective parent.

Bill of materials structure a listing of bill of materials for one unit of a complete product that shows the organization of the bills by level or stage of production required to complete the unit of finished product.

Continuous flow production a lotless production where production flows continuously rather than being proportioned into lots.

Cycle counting an inventory audit technique used to detect errors in inventory quantities.

Cycle time the period of time between successive completions of work activities, jobs, subassemblies, or finished products.

Cycle time/crew size matrix a listing of crew sizes necessary to complete a given amount of demand in a fixed period of time. The crew sizes are determined by dividing the cycle time required to complete demand into a given unit's labor content time requirement.

Deduct list a listing of component parts, like a bill of materials, that will have to be deducted from inventory for each unit of the parent part or subassembly manufactured. A deduct list is required for each deduct point that makes up the JIT production system and is a list of component parts that should be deducted from inventory in a stock area for each unit of its parent part.

Deduct point a point in the production process where inventory transactions are recorded. A deduct point can be at a storage area, work station, or any other location in a facility. The inventory transaction is usually a reduction of inventory of the items on the deduct list for that deduct point.

Exception report an MRP report that lists or flags only those items which deviate from an expected plan.

Flat BOM file a single product completion level BOM file. This file usually lists total component parts rather than including pseudo assemblies or pseudo component parts whose purpose is chiefly to denote activities at specific production levels or stages.

Intermittent manufacturing a type of manufacturing organization in which productive resources are organized according to function (e.g., painting department, assembly department, etc.). Usually the job orders pass through the functional departments in lot-size quantities. For small or individual unit lot sizes, this type of operation is sometimes called a job shop.

Labor content the required labor time to complete one unit of inventory in a given or fixed manufacturing process. Labor content is found for a given lot size by multiplying the number of crew members used to produce the lot size order by the total time required for each crew member to complete a lot-size order, then dividing this product by the number of units completed.

Line balancing procedure a heuristic procedure by which the elemental tasks that make up a job are allocated to workers or work stations along an assembly line or in a continuous flow production system.

Load report an MRP report that compares existing production capacity with future use of that capacity. Load reports can be expressed in hours of labor or units of product for work stations, departments, or even individual workers.

Manufacturing resource planning (MRP II) a computer-based system that is used to plan and control the use of inventory in the manufacture of finished goods while sharing a common data base with computer systems in accounting and marketing.

Master production schedule (MPS) a broad scale schedule that establishes unit production requirements for all products over a fixed time planning horizon. The MPS sets the daily, weekly, or monthly production in units for an entire production facility. The MPS is used as a plan to guide an entire operation toward specific unit production goals.

Material requirements planning (MRP) a software system that is used to help plan and control the use of raw materials, subassemblies, and component parts. An MRP system establishes the usage of various inventories over a specific planning horizon.

Net-change MRP a type of MRP system that implements changes in a given MRP schedule. Since a change in a single lead time may only impact on a single product, this system only changes a portion of a given MRP schedule as opposed to completely revising the schedule.

On-line an expression used to describe a computer-based transaction where one is in a current state of interacting with a computer.

Parent a subassembly, component part, or finished product that is on the next "product completion level" in the "product structure" file of the BOM.

Planned order release a suggested MRP schedule of manufacturing orders or purchase orders. These are suggested order quantities, release dates, and due dates for all inventory items under the control of the MRP system.

Product completion level a level or stage of production requiring some type of production activity.

Pseudo bill a listing of an artificial grouping of components or production activities in a bill of materials format to express a production activity at a specific level or stage of product completion. They can also be called "pseudo assemblies" or "pseudo components."

Push system a system of using raw materials and component parts inventory to meet a schedule based in part on forecast demand of the finish product. Inventory is pushed through the production system to meet the forecast demand regardless of the actual demand for the finished product.

Rate-based master production schedule an MPS that is based on a daily rate of production. This type of MPS is usually flexible to take into account the need for changes in production rates that are necessary to respond to daily demand requirement fluctuations.

Regenerate MRP a type of MRP system that regenerates an entire MRP schedule for any changes in the data input of the system.

Repetitive manufacturing a type of production operation where discrete units of a fairly homogenous product are produced at relatively high speeds and volumes, also characterized as continuous flow production in a sequentially organized operation.

Requirements explosion a process of calculating the demand for component parts of a parent item by multiplying the parent item requirements by component usage in bill of materials.

Rough-cut capacity planning the process of converting the master production schedule into capacity needs for key resources such as human resources, machinery, and money.

Stock area a place in a manufacturing facility where inventory stock is temporarily stored and where inventory is consumed. A stock area may be a place along a production line, a work center, etc.

Synchronized production a type of continuous flow production used to describe JIT production.

Time bucket a period of time (usually a week for MRP systems) used to define a set of production activities or inventory requirements.

Time-phased an MRP technique of expressing future demand, supply, and inventories by time period.

Uniform daily production a production scheduling approach that seeks the same production rate for each product produced every day. Typically used in JIT operations, this approach allows the typical monthly MPS rate of production to be sliced into daily adjustable production rates to be more flexible to meet shifts in demand.

Work center a specific production facility usually consisting of one or more workers, robots, or machines.

Work-in-process (WIP) inventory in various levels or stages of production completeness. This would include inventory from raw materials up to a finished product.

Discussion Questions

1. How is backflushing in a computer-supported JIT different from a requirements explosion in MRP II?

2. Why are stock areas needed in an MRP II/JIT system?

3. Explain the difference between a multi-level and single-level BOM file. Where and why do we need both types of BOM?

4. Explain the purpose of deduct points and deduct lists. How do they support decision making in an MRP II/JIT system?

5. How is rate-based MPS different from any other MPS?

6. How does a rate-based MPS create flexibility in an MRP II/JIT system?

7. Why is flexibility in purchasing policies necessary for the MRP II/JIT system to operate? How should the policies be flexible?

8. Why is flexibility in worker job assignments necessary for the MRP II/JIT system to operate? How should management accomplish this flexibility?

9. Why do we need a synchronized facility layout for the MRP II/JIT system to operate? How is it different from a departmentalized layout?

10. Why is an alteration in customer ordering policies necessary for the MRP II/JIT system to operate? How should the policies be altered?

Problems

1. Redraw the multi-level BOM structure below into an acceptable single-level structure for a computer-supported JIT system.

2. Redraw the multi-level BOM structure below into an acceptable single-level structure for a computer-supported JIT system.

3. Redraw the multi-level BOM structure below into an acceptable single-level structure for a computer-supported JIT system.

4. Given the stock area deduct points and deduct lists below, determine the number of units of component A that will be used in the completion of ten units of product X. How many units of B?

5. Given the stock area deduct points and deduct lists below, determine the number of units of component A that will be used in the completion of fifteen units of product X. How many units of B? How many units of C? How many times was the deduct point SA2 activated to deduct its list of parented parts in the manufacture of fifteen units of product X?

6. In the rate-based MPS below, what would the production rate have to be set at in February 1993 to meet the demand in that month?

Production/Inventory Item	Month Jan/1993	Feb/1993
Number of working days	22	20
Production rate (units/day)	100	?
Backlog of units (for month)	120	0
Unit forecast (for month)	2.000	2.322
Total units (for month)	2,120	2,322
Production (units/month)	2,200	?
Beginning inventory (for month)	98	178
Available inventory (for month)	2,298	?
Ending inventory (for month)	178	?

7. In the rate-based MPS below, what would the production rate have to be set at in February 1993 to have zero ending inventory at the end of the month?

Production/Inventory Item	Month Jan/1993	Feb/1993
Number of working days	22	20
Production rate (units/day)	120	?
Backlog of units (for month)	500	100
Unit forecast (for month)	2.640	2.700
Total units (for month)	3,140	2,800
Production (units/month)	2,640	?
Beginning inventory (for month)	200	300
Available inventory (for month)	2,840	?
Ending inventory (for month)	300	0

8. In the rate-based MPS below, what would the production rate have to be set at in February 1993 to have zero ending inventory at the end of the month?

	Month	
Production/Inventory Item	*Jan/1993*	*Feb/1993*
Number of working days	22	20
Production rate (units/day)	200	?
Backlog of units (for month)	900	300
Unit forecast (for month)	4,000	3,000
Total units (for month)	4,900	3,300
Production (units/month)	4,400	?
Beginning inventory (for month)	200	0
Available inventory (for month)	4,600	?
Ending inventory (for month)	0	0
Shortage backordered to next month	300	?

9. If the cycle time for a product is ten minutes, based on 100 units of demand, and its labor content is 100 minutes, what would the crew size of the computer-supported JIT operation have to be to meet a demand of 100 units?

10. Assuming a five-day work week of eight hours each day, what is the cycle time for a product if 200 units of the product are produced in two weeks? If the labor content of each unit is 12,000, what size crew will be needed to staff a computer-supported JIT work center of one worker each?

References

[1] W. J. Stevenson, *Production/Operations Management*, 3rd ed., Irwin, Homewood, IL, 1990, Chapter 11.

[2] T. E. Vollmann, W. L. Berry, and D. C. Whybark, *Manufacturing, Planning and Control Systems*, 2nd ed., Irwin, Homewood, IL, 1988.

[3] R. Discenza and F. R. McFadden, "The Integration of MRP II and JIT Through Software Unification," *Production and Inventory Management Journal*, Vol. 29, No. 4, 1988, pp. 49–53.

[4] Hewlett-Packard Corporation, "HP JIT: Just-in-Time Manufacturing," from *HP Manufacturing Management II: General Information Manual*, Corporate Information Systems Center, Palo Alto, CA, 1985, Section 11, pp. 1–7.

[5] J. W. Toomey, "Establishing Inventory Control Options for Just-in-Time Applications," *Production and Inventory Management Journal*, Vol. 30, No. 4, 1989, pp. 13–15.

[6] R. J. Schonberger, *World Class Manufacturing*, The Free Press, New York, 1986.

[7] L. Heiko, "A Simple Framework For Understanding JIT," *Production and Inventory Management Journal*, Vol. 30, No. 4, 1989, pp. 61–63.

[8] A. Rao, "Manufacturing Systems—Changing to Support JIT," *Production and Inventory Management Journal*, Vol. 30, No. 2, 1989, pp. 18–21.

[9] F. W. Hazeltine and R. J. Baragallo, "The Key to Competitive Success in the 1990s," *Production and Inventory Review*, Vol. 10, No. 2, 1990, pp. 41–46.

[10] K. A. Wantuck, *Just In Time for America*, The Forum, Ltd., Milwaukee, WI, 1989, 230–232.

[11] International Business Machines, "COPICS Inventory Management," from *IBM COPICS Manual*, International Business Machines Corporation, White Plains, NY, 1989.

CASE 6-1

A JIT Waterbed Nightmare

The Boston Waterbed Company (BWC) of St. Louis, Missouri, produces waterbeds for its chain of retail stores located throughout North America. They also produce units for many of the large department store chains throughout the United States. Their plant, located in Overland, Missouri, is a dedicated plant to a single model of waterbed, called the "float or sink" model or "floink" for short. The multi-level BOM file for this product is presented in Exhibit 6-1. As we can see in Exhibit 6-1, the floink model waterbed consisted of only a handful of raw materials and component parts. The company currently operates in an intermittent batch mode of production with an MRP II system to assist in management's planning and control activities. Currently, monthly lot-size production is fixed at 10,000 units of the model. Each quarter the changes necessitated by exceptions to the MRP-based plan are implemented and a new MRP MPS is generated using a regenerate MRP system. Management of BWC felt the current planning system needed to be more flexible to meet shifts in monthly demand.

The company received all of its raw materials from a local distributor and component parts from several regional suppliers. All of the inventory items were purchased in large quantities to take advantage of quantity discounts. As can be seen in Exhibit 6-2, the large order lot sizes require substantial inventory space in the plant.

The company's Overland plant consists of six departments, arranged basically in order of their sequence of operations according to the BOM file. Each department's staff is only trained to perform the necessary department production activities. Cross-training within the department is encouraged, but no cross-training of activities between departments is permitted because of a labor agreement.

The company knows the conversion to a computer-supported JIT operation will be complicated by carry-over problems still facing the MRP II system. For example, the current backorders of product for which the firm did not have capacity during the current planning period have to be included in the new JIT-based master production schedule. The backlog in January (the first month in the new planning horizon) is 300 units and the backlog in February is 200 units. These backordered units must be added to the production requirements in the future planning period.

EXHIBIT 6-1 • *A Multi-level MRP Manufacturing Bill of Materials Structure File for Product "Floink"*

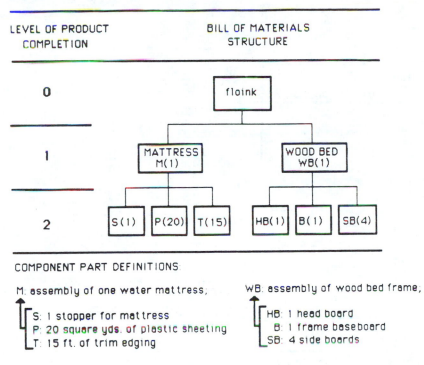

LEVEL OF PRODUCT
COMPLETION

BILL OF MATERIALS
STRUCTURE

COMPONENT PART DEFINITIONS:

M: assembly of one water mattress;
S: 1 stopper for mattress
P: 20 square yds. of plastic sheeting
T: 15 ft. of trim edging

WB: assembly of wood bed frame;
HB: 1 head board
B: 1 frame baseboard
SB: 4 side boards

To overcome the inflexibility of the existing planning and control system, the management of BWC decided to convert their MRP II controlled operation in Overland into a joint MRP II/JIT operation. To accomplish this conversion they have called you in as a consultant to advise them.

Case Questions

1. What changes will have to be made in the current plant operation to implement the conversion of the MRP II system to a MRP II/JIT system? Prepare a step-wise plan to change BWC's data input requirements of the system, physical layout of the facility, and scheduling. Be specific to cover any areas where management will have to alter current operating policies or negotiate a new arrangement with individuals in the operation of the plant. Define what must be in place in order to make the new JIT-based system successful.

EXHIBIT 6-2 • *Overland Plant Layout and Department Activities*

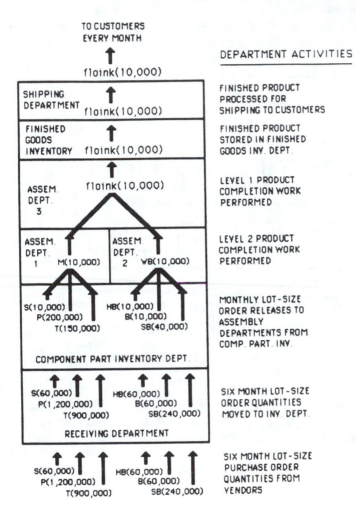

DEPARTMENT ACTIVITIES

2. What is the single-level BOM structure file for the multi-level BOM structure file in Exhibit 6-1? Draw the new BOM labeling all inventory items.

3. Assuming the number of working days in January is 22 and in February is 20, the daily production rate in January is 500, January's beginning inventory is 150, its unit forecast is 10,000, and February's unit forecast is 13,000 units, what would February's daily production rate have to be set at in order to achieve little or no ending inventory in February? Show your work by preparing a rate-based MPS like the one presented in Table 6-1.

4. The company will obviously face the crew size decision in future periods after the implementation of JIT in the Overland plant. Based on a monthly production rate of 10,000 units in 20 days, what is the necessary cycle time for the new JIT facility? If a crew of five workers will achieve the 10,000-unit monthly production level, how large must the crew be to satisfy a 13,000-unit production level? A 15,000-unit production level? A 20,000-unit production level?

CASE 6-2

JIT Learning

The Holton Manufacturing Company (HMC) is a major manufacturer of furniture parts and accessories. They specialize in items like door knobs, drawer handles, etc. The company manufactures over 3,000 different products. The family of door handle products alone represented a total mix of 54 different products. Door handles represented about 30 percent of sales and about 50 percent of the company's profits. The company is very successful in generating a low-cost product to customers, while at the same time generating very high profits through high volume sales for the company.

HMC obtains some of their items from vendors and fabricates the rest from sheet metal in their own plants. HMC operates several plants whose production activities are tied together by a central computer system. The primary production scheduling and inventory order activities are performed by MRP II system software.

Demand for all door handle products has continually grown at a fairly constant rate of 4 percent per year for several years. Indeed, the demand for the products has shifted from a demand push environment of MRP to a demand pull situation as the known demand exceeds the company's production capacity. HMC can sell every door handle they can produce. HMC felt that their door handle product family demand had become substantial enough to warrant a "focused factory" dedicated to just door handle production.

In planning the focused factory the management of HMC looked for possible problem areas before initiating the move. The vendor-acquired parts for the product family are separate from any of their other products and posed no constraints. It would be easy to have focused vendors for the plant. The equipment at their candidate factory can support the door handle product family production requirements, and there is a sufficient support of labor to complete the necessary work. The plant's physical layout is currently structured in ten equal sized rooms, arranged in a 2 by 5 grid. The current layout supports the MRP type of batch system that the company has been using for years. The layout can be changed if needed.

To ensure the success of the new focused factory, the management of HMC hired a management consultant from outside the organization to head the program of moving. It was felt that the consultant could not only advise on the move to the new plant, but if the new operation required any redesigning efforts to improve production planning and control, the consultant's expertise would be beneficial. After a few days of analysis, the consultant suggested that the focused factory and the necessary changes to make it work provided HMC an opportunity to experiment with the introduction of a combined MRP II/JIT type of operation. The consultant felt that the new factory should focus on JIT principles that could operate within the MRP II computer-based planning environment.

Case Questions

1. Suppose you are the consultant and must prepare a lecture that will seek to sell HMC management on how the MRP II/JIT system is better than the MRP II system alone they currently use. What would you put in such a lecture? Outline the lecture you would present.

2. If the typical door handle product consists of eight component parts, how would you redesign a flow shop layout that would permit JIT to operate in the MRP II environment? Draw a diagram like the one in Figure 6-8.

3. From the description of the candidate factory and HMC's products, what additional possible problem areas might inhibit the implementation of JIT in this new factory?

CHAPTER SEVEN

Integrating JIT in Service and Administrative Systems

Chapter Outline:

Learning Objectives

After completing this chapter, you should be able to:

1. Explain what a service system is and how it may be a part of a manufacturing operation.
2. Describe different criteria used to characterize service systems.
3. Explain how JIT principles can be integrated into different types of service systems.
4. Explain how an expert system can help integrate JIT principles into service systems.
5. Explain several rationales for implementing JIT principles in administrative systems.
6. Explain several strategies and their related tactics for integrating JIT principles in administrative systems.

Introduction

Service systems can be defined in many ways. In businesses, **service systems** seek to perform economic activities that produce an intangible product that adds value or creates utility. Within manufacturing operations service systems can exist that facilitate the production of the tangible products from manufacturing systems. An example of this type of service system is an accounting department (the service system) providing cost information (the intangible service product) for manufacturing activities. Some businesses that are considered service also provide customers with tangible **goods** as a part of the service. The fact that a service system provides tangible goods as a part of their service product does not disqualify them as service systems. In the hospital business, for example, patients are receiving the intangible service of health care from doctors and nurses, along with tangible goods like medicine or bandages. The convenience of the delivery of tangible goods is by itself considered a service product of the service system that delivers it. There are also whole organizations that can be classified as service systems. A motel for travelers is one example. With the exception of food, the motel's product of a good night's sleep is an entirely intangible service. All business organizations are made up in part or in whole of service systems. In this chapter we will use the term "service system" to describe either a service system within a manufacturing operation or an entire service operation [1, 2, 3].

In the early 1990s, more than seven out of ten U.S. workers are employed in what the U.S. Census Bureau calls the **service sector.** Continued growth in the service sector is expected through and beyond the year 2000 by the U.S.

government. This dominance of service systems in the U.S. economy makes it clear that the greatest area of possible application, now and in the future, for philosophies like JIT is in service systems. Applications of JIT in service systems are beginning to appear in the literature [4, 5]. It is essential to understand how JIT can be integrated into the diverse set of industries that use service systems.

The purpose of this chapter is to examine the integration of JIT with service systems, and to explain how JIT principles can be used in service systems. This chapter will begin by relating applicable JIT principles to criteria used to describe various service systems. The unique nature of each service system prohibits any direct JIT principle to be consistently applied in all types of a specific service industry. By explaining how various JIT principles apply to criteria that characteristically represent service systems, this chapter seeks to enhance the generalized applicability and integration of JIT into all types of service systems. To accomplish this explanation, a series of twelve types of service organizations will be selected to represent the twelve types of service criteria that will be discussed. A discussion of how each service organization can support JIT principles will be presented. In addition, this chapter will also present a discussion of a new technology that can be used to help integrate service systems with JIT principles.

One special case of service systems is the administrative services that managers in all types of organizations perform. **Administrative services** are tasks that are performed by workers or managers that indirectly support the production of an organization's service or manufactured products. Administrative services are those activities that are required to complete business transactions, but not directly related to the production of an organization's product. These services are required in all types of organizations (business, government, schools, etc.) and in all departmental areas that make up organizations, including accounting, finance, marketing, and operations. In accounting departments, administrative services can include estimating product costs, preparing income and balance sheets, or performing audits. In finance departments, administrative services can include processing loan applications, performing equipment investment analyses, or developing a long-term capital investment plan report. In marketing departments, administrative services can include order taking, developing price quotes or developing sales forecasts. In operations, administrative services can include picking orders, preparing production schedules, and training personnel. Administrative services are provided by administrative systems. An **administrative system** is a set of procedures, guidelines, rules, or policies that provide the method by which an administrative service product is to be prepared and/or delivered.

Another purpose of this chapter is to examine the integration of JIT with administrative systems. A section in the chapter is devoted to a discussion of general strategies and tactics for integrating JIT principles. In addition, this chapter will also describe commonly used methods that can facilitate the implementation of JIT principles in administrative areas.

Integrating JIT in Service Systems

A variety of criteria are used to identify what comprises a service system. Twelve of the more common of these service system criteria include [3, p. 27]:

1. Service systems produce an intangible product.
2. Service systems produce variable, nonstandard output.
3. Service system output is perishable and cannot be carried in inventory.
4. Service system processes require a high degree of customer contact.
5. Service systems require customer participation in generating output.
6. Service system skills are sold directly to customers.
7. Service system output cannot be mass produced.
8. Service system workers exercise a high degree of personal judgment in performing their work assignments.
9. Service systems are labor intensive.
10. Service system facilities are decentralized and located near customers.
11. Service system measures of effectiveness are subjective.
12. Service system quality control is primarily limited to process control.

All service systems exhibit at least some of these criteria in the performance of their various functions. Collectively, they represent descriptive criteria or operating characteristics that are common in service systems. JIT principles can support, either directly or indirectly, all of these operating characteristics. The extent to which JIT principles can be applied to these criteria offers opportunities for JIT integration in service systems. Let's examine each of these twelve criteria, and see how JIT principles have or can be applied to them. Since the variety of service systems is almost endless, a basis for limiting the discussion on possible JIT applications is necessary. For each of the twelve criteria, only one specific type of service operation will be described. The selection of the service system will be based on how well that operation typifies the particular service system criteria.

1. Service Systems Produce an Intangible Product A medical doctor giving a diagnosis or providing a chemical therapy is an example of a service system (i.e., the doctor) providing an intangible service product (i.e., medical care). JIT principles can be applied in a variety of ways to this type of operation. Of the JIT inventory management principles, some of the most applicable to this type of service system and characteristic include cutting purchasing costs, improving material handling, seeking zero inventory, and seeking reliable suppliers. Doctors need an inventory of drugs, syringes, etc., and they need them in time to do the job. The risk of not having the necessary drug may be serious in rural areas of the U.S. and may necessitate some inventory supplies for convenience. For a doctor located in a major U.S. city with a drug store around every corner and a drug vendor a call away, the JIT goal of zero inventory is an applicable principle to follow. While exceptions for medicine that treats life-threatening problems can be made, doctors who can cut back on highly perishable medicine can cut purchasing costs without sacrificing service. The reduced inventory will necessitate seeking reliable suppliers and reducing their number as per JIT purchasing principles. This supplier reliability will free the doctor's and staff workers' time for more productive efforts.

Of the JIT production management principles, some of the most applicable to this type of service system and characteristic include seeking a synchronized pull system, using automation where practical, seeking improved flexibility in workers, and improving communication and visual control. Appointment systems used by doctors in their own offices can be used as JIT **pull systems,** since non-emergency patients can be scheduled to coordinate with medical inventory arrivals, the use of rented equipment, and the doctor's other medical duties. JIT doctors, like JIT manufacturers, must make investments in automated equipment that will save time and money and add value to their customer service. Doctors should be guided by the JIT principle of automation in equipment acquisitions to guard against the loss of processing flexibility, and balance automation benefits. Lab technicians are far more flexible than any existing computer-based medical equipment in running medical tests, while the equipment offers processing speed and accuracy (i.e., quality). Doctors should seek office staff with the highest degree of skills, as suggested by JIT principles, to permit the greatest degree of flexibility in assignments. This is particularly true in small office operations where vacations and untimely staff leaves can be crippling to an office's operation. It can also be crippling to the doctor's clientele when the leaves reduce the flow of delivery of service to desperate patients who consider "doctor reliability" an important part of the service product. Doctors should establish a period of once a week or at least once a month to talk over operation problems with all staff members. The JIT principle of letting workers suggest improvements can be taken advantage of in doctors' offices as well as in manufacturing plants. Doctors should set their offices up with visible directions (signs with arrows pointing in the right direction) to minimize patient confusion in moving through an office suite, and maximize patient flow. This will save staff and doctor time in giving patients directions, and improve the productivity of the operation.

Of the JIT quality management principles, some of the most applicable to this type of service system and characteristic include: maintaining process control and making quality everybody's responsibility, giving the workers authority to share in the control of product quality, maintaining a 100 percent quality inspection of products, and seeking continual quality improvement. Doctors have to be champions for quality in their office operations and in their hospitals. Health care is an industry that has some of the highest quality standards. Characteristic of one of the medical practices in hospitals is a periodic meeting where doctors and sometimes staff meet to discuss the progress or lack of progress with patients they are servicing. The purpose of such meetings is to allow a shared basis of quality control on patient services, and offer suggestions on problem patients. This type of meeting serves like a combined process control system and quality control circle for the hospital. Indeed, doctors have for years prepared and posted at a patient's bed or nurse station "process control charts" to monitor and communicate a patient's progress. Hospital operating rooms have also used the JIT principle of a 100 percent inspection. As a basic operating room rule, operations are witnessed by at least one doctor whose job it is to observe (and advise in the case of defective behavior). Doctors are also characterized as continually seeking improvement in the quality of the work they perform, as reflected in the rapid changing pharmaceutical and medical equipment industries that support and supply doctors.

2. Service Systems Produce Variable, Nonstandard Output A management consultant giving advice on the implementation of some system or the acquisition of some equipment is an example of a service system (the consultant) providing an intangible service product (expert advice). A consultant's information service output is different or nonstandard for each job he or she does, because clients' problems are unique. Regardless, JIT principles can still be applied in a variety of ways to this type of operation. Of the JIT inventory management principles, one of the most applicable to this type of service system and characteristic is to cut lot sizes and increase the frequency of orders. The intangible nature of "advice" cannot be stored or inventoried, but the materials or goods used to impart advice (educational books, pamphlets, handouts, etc.) can be treated in the same way with JIT as we might treat manufacturing materials. Rather than running off large volumes of these support materials, consultants should supply them to users on a demand pull basis. The additional costs of having small orders run off for individual presentations might be compensated for by avoiding the risk of portions of larger runs becoming outdated before they are used.

Of the JIT production management principles, some of the most applicable to this type of service system and characteristic include: seeking production scheduling flexibility, seeking a synchronized pull system, seeking improved flexibility in workers, cutting production setup costs, using automation where practical, and improving communication and visual control. Consulting work requires scheduling flexibility to accommodate the more urgent demands of some clients over others. It is completely a demand pull situation, since clients have to hire consultants before the work can begin. Consulting work requires a great deal of flexibility in the consultant skill and knowledge levels, because of the wide range of products (solutions to client problems) they have to be able to offer. Consultants can cut their setup time for advisory presentations or meetings by planning group meetings with larger numbers of clients of a firm. They can also enhance their communication with clients by using advanced video technology and graphics. These JIT practices save consultants' time and effort, providing an improved product to the customer.

Of the JIT quality management principles, some of the most applicable to this type of service system and characteristic include: making quality everybody's responsibility, giving the workers authority to share in the control of product quality, requiring self-correction of worker-generated defects, maintaining a 100 percent quality inspection of products, and seeking continual quality improvement. A consultant's presentation of advice is usually a shared product that combines the intelligence and the skills of technicians, including those that will help prepare the materials for the presentation to customers. All participants should share the responsibility and the authority to make the finished product of the highest quality possible. The consulting advice should be checked at each stage of development, ensuring the JIT 100 percent quality inspection. A consulting firm, whose reputation is only as good as the results of their last job, should seek continual product quality improvements by encouraging staff and other personnel to continually keep up to date in the latest in technology and management methods.

3. Service System Output is Perishable and Cannot be Carried in Inventory   A fresh fruit vendor selling fruit in an open market is an example of a service system (the vendor) providing an intangible service product (convenience of location). The fruit that is being sold is highly perishable. JIT principles can be applied in a variety of ways to this type of operation. Of the JIT inventory management principles, some of the most applicable to this type of service system and characteristic include cutting lot sizes and increasing frequency of orders, improving material handling, and seeking reliable suppliers. Fresh fruit has to be fresh and available in time for its consumption. Unfortunately from a JIT point of view, the presence of the fruit in a fruit store is a major determiner of the sale, and as such, some inventory is necessary. While the JIT zero inventory principle may not be possible, a reduction in lot size is always possible. Vendors should achieve a good relationship with suppliers and establish long-term agreements permitting more frequent, but smaller order quantities for fruit to minimize spoilage risks. The vendor's contract with the supplier should also be structured to permit some flexibility to allow for changes in day-to-day market fluctuations. Vendors should also seek to improve their material handling practices to avoid damage to their products. Selling fruit by the crate, having the supplier down-size the packages of fruit, and standardizing the crates to fit the counter areas will help eliminate or avoid material handling efforts.

Of the JIT production management principles, two of the most applicable to this type of service system and characteristic include: seeking a synchronized pull system and improving communication and visual control. By establishing a known customer market, a fruit vendor can more accurately anticipate demand and seek to achieve an almost synchronized demand pull type of operation. The layout of the market should be structured to facilitate a customer "flow shop" operation. Perhaps an application of GT cells is appropriate, where one cell is for vegetables, one cell is for fruit, etc. A sufficient number of alterable signs with flexibility for labeling prices, products, and quantities should be structured into the layout to achieve desired JIT "seiton" objectives.

Of the JIT quality management principles, some of the most applicable to this type of service system and characteristic include maintaining process control, making quality everybody's responsibility, giving the workers authority to share in the control of product quality, requiring workers to perform routine maintenance and house cleaning duties, and seeking a long-term commitment to quality control efforts. Process control charts can be used to keep track of the units of fruit that suppliers send that can't be sold because of defects like bruises caused by the suppliers' poor handling. Also, every worker should be held responsible for inspecting and adjusting fruit stands to promote the appearance of quality to the customer. As in JIT manufacturing operations, these inspections can be performed with little or no cost to productivity by having the workers do them as they pass in the performance of their other duties, including routine maintenance of scales, and keeping the market clean. All of these JIT activities will help build worker pride, and help to motivate quality as a long-term habit.

4. Service System Processes Require a Hgh Degree of Customer Contact A fast-food vendor is an example of a service system (the vendor) providing an intangible service product (food preparation). An intensive effort in preparing, packaging, and delivering fast food requires a high degree of customer contact. JIT principles can be applied in a variety of ways to this type of operation. All of the JIT inventory management principles are applicable to this type of service system and characteristic, including cutting lot sizes and increasing frequency of orders, cutting buffer inventory, cutting purchasing costs, improving material handling, seeking zero inventory, and seeking reliable suppliers. Fast food vendors order food materials, napkins, spoons, etc. from suppliers. Consistent with JIT principles, they should experiment with cutting lot size orders to see how close they can come to a zero inventory goal by balancing the cost of missed sales with the cost of saving on-hand inventories. Obviously, the nature of "fast food" means "fast," and some buffer inventory will be necessary. Those vendors who order a month's or a week's supply of an inventory item at a time should try cutting down the amount, and place smaller, more frequent orders. This process of "stressing the system" with less inventory starts the JIT benefits of reduced inventory and improvements in production processes.

Of the JIT production management principles, some of the most applicable to this type of service system and characteristic include: seeking production scheduling flexibility, using automation where practical, seeking improved flexibility in workers, cutting production lot size and setup costs, allowing workers to determine production flow, and improving communication and visual control. The demand for fast food is not uniform on a daily basis, and is difficult even when scheduling on an hourly basis. The best and most common response in scheduling demand is to provide JIT flexibility in scheduling the production resources (the workers) to meet the demand. Investments in automation in fast food operations can be helpful when they focus on production timing and seeking to provide visual or voice warning systems to motivate worker compliance (e.g., food timers or temperature gauges for ovens). Management should also encourage the development of JIT reduced lot size production during differing periods of the day to minimize spoilage. Filling an oven with 50 burgers when only 10 will be ordered in the next 20 minutes causes waste and poor quality.

Of the JIT quality management principles, some of the most applicable to this type of service system and characteristic include: making quality everybody's responsibility, seeking a high level of visibility management on quality, maintaining strict product quality control compliance, giving the workers authority to share in the control of product quality, requiring self-correction of worker-generated defects, requiring workers to perform routine maintenance and house cleaning duties, seeking continual quality improvement, and seeking a long-term commitment to quality control efforts. All workers in a fast-food operation must be responsible for product quality and given the authority to suggest and implement improvements in cooking, maintenance, cleaning, etc. that will eventually improve product quality. Fast food managers must continually remind workers of quality standards (this is particularly important with the high turnover these operations experience), and enforce strict compliance by making workers who violate standards correct customer orders. Redoing orders in front of impatient customers will be a lesson that will help make quality a habit. Cleanliness duties in work areas and the facility as whole should be specifically defined, monitored, and where exceptional work is observed, rewarded as a long-term effort to show management's commitment to quality in the workplace.

5. Service Systems Require Customer Participation in Generating Output   An amusement park that provides games and rides is an example of a service system (the park) providing an intangible service product (entertainment). All of the games and most rides in an amusement park require customers to interact and participate in the delivery of the entertainment product. Their participation simply means that we have to enlist their aid in applying some of the JIT principles. Of the JIT inventory management principles, some of the most applicable to this type of service system and characteristic include: cutting lot sizes and increasing frequency of orders, cutting purchasing costs, improving material handling, and seeking reliable suppliers. Novelty items given as prizes for the games, spare parts for the machinery of the rides, food, and drinks are all inventory items or goods that managers should seek to order in smaller lot sizes and order in greater frequency to start the JIT benefits of reduced inventory. Cutting purchasing costs and seeking reliable suppliers through research and negotiation of long-term contracts are sound JIT principles to achieve improved supplier service and improved operations efficiency by saving order processing time. The layout of the amusement park should also be structured to facilitate movement of customers through the park, and to enable customers to switch from rides with long lines to ones with shorter lines more easily. Customers are motivated to help implement JIT principles, like improving material handling, because they represent a type of material flowing through the production process (the amusement park), and they seek to minimize their own waste of time (by avoiding the wait in long lines).

Of the JIT production management principles, some of the most applicable to this type of service system and characteristic include: seeking uniform daily production scheduling, seeking production scheduling flexibility, seeking a synchronized pull system, seeking improved flexibility in workers, allowing workers to determine production flow, and improving communication and visual control. The customer demand in an amusement park operation defines production scheduling. By selectively advertising and promoting differing park rides and games, the demand, and therefore the production scheduling activities, can be brought into alignment for more uniform daily demand and improved resource planning efforts (for instance, allocating park staff workers to popular advertised rides and games). The coordination and timing of rides can also be used by park management to improve customer flow. By structuring long rides to begin at specific times, customers can plan to make better use of their time, and enjoy other shorter rides and game activities between starts of long rides. Parks can also coordinate ride timing so riders are let out of one ride in a location where another conveniently located ride is just starting up. This JIT principle of delivering customers who can participate just-in-time for the next ride will keep customers on rides, maximizing their entertainment benefits and minimizing the waste of waiting in lines. Amusement parks should also seek to hire staff workers who possess a wide range of skills to cover the diversity of applications found in today's technologically complex and highly diversified entertainment parks. Staff workers should also have decentralized control in operating machinery which is not monitored on a per-use basis (no 100 percent inspection possible), and yet could be dangerous to customers. The staff workers play a key role in ensuring customer safety on rides and in helping customers to use rides to maximum satisfaction. JIT principles of automating are a practical support for the worker's role in that they seek to guide parks to structure their layout to permit an ease of management control. Many parks use remote video equipment for security and trouble shooting. Managers and staff workers should be able to access this technology quickly to communicate problems and receive instructions via personal electronic communication systems.

Of the JIT quality management principles, some of the most applicable to this type of service system and characteristic include: making quality everybody's responsibility, seeking a high level of visibility management on quality, giving the authority to share in the control of product quality, requiring self-correction of worker-generated defects, requiring workers to perform housecleaning duties, seeking continual quality improvement, and seeking a long-term commitment to quality control efforts. Inviting customers, staff workers, and managers to comment on the amusement park entertainment products and offer suggestions for improving service not only obtains useful ideas, but reinforces the fact that the park is continually interested now and in the future in improving all product quality. All members of the park's staff should be given the responsibility and authority to share in quality control activities. Instituting quality control circles consisting of staff workers, managers, and mechanical engineers (i.e., the technicians who build the rides and games) might be a useful strategy to deal with problems reported, and show a park's shared commitment to quality. JIT parks should encourage staff workers to take on housecleaning duties on rides, park grounds, and game stands they operate. These duties not only enhance the entertainment value for the customer in maintaining a clean area, but develop discipline about quality, making it a habit.

6. Service System Skills are Sold Directly to Customers A city civil servant (city mayor, police officer, etc.) providing service to the community is an example of a service system providing intangible service products and goods (e.g., governmental order and successful delivery of city services). In a government, what the workers or administrators do determines the product the people (the customers) receive. The customers basically have to accept the skills of government administrators as long as they hold office because of the monopolistic nature of governmental service systems (e.g., getting a ticket from a police officer). JIT principles can be applied in a variety of ways to this type of operation. All of the JIT inventory management principles are applicable to this type of service system and characteristic. JIT city governments can and have benefited from the application of JIT principles to the various material goods that cities deliver. Electricity is an example of a type of service "good" that can't be stored, but must be delivered just-in-time for its use to the public. Timing its synchronized generation from power plants or purchase from other sources of generation is a perfect past example of governments having implemented the JIT inventory management principle of zero inventory for decades. Also, city governments purchase all types of materials to which JIT inventory principles could be applied. For example, cities purchase large quantities of paper for many publications. They could cut paper lot sizes and increase frequency of orders, cut buffer inventory, cut purchasing costs, and seek reliable suppliers through long-term contracts, and improve material handling to minimize waste.

All of the JIT production management principles are applicable to this type of service system and characteristic. In city governments, as in JIT operations, the workers chiefly determine work flow, regardless of the customer demand. While it is not usually done for hiring reasons or because of the specialized departmentalization of the government, city governments should try to embrace the JIT principle of improved flexibility in workers to enhance customer service in poorer economic times of financial cutbacks. Some city governments do use JIT synchronized scheduling flow ideas. For example, by coordinating the lighting pattern of city stop lights, the timing of go lights can be set to allow cars to flow in one direction with almost no stopping. Many city governments have used this method during morning and late afternoon rush times when workers drive into and out of city areas. This helps to minimize travel time by eliminating wasted idle time of drivers waiting at poorly timed lights. Many cities are also automating their highway systems with electronic sensors that monitor traffic congestion and help control the lights letting traffic onto highways. Cutting the lot sizes of flow onto the highway improves flow and reduces traffic jams (i.e., bottlenecks in the production system). Emergency phone systems such as the use of 911 also help improve communication and efficiently obtain city services for critical areas requiring a solution, thereby focusing problem solving resources, like JIT, toward problems that are disruptive to the city system.

All of the JIT quality management principles are applicable to this type of service system and characteristic, but are not always applied. JIT city governments can use process control charting in monitoring crime, and they can make everybody in government responsible for quality. While most politicians promise a high level of visual management of quality, many of them should be forced to embrace the other JIT principle of maintaining strict product quality control compliance. City government rarely gives workers the authority to share in the control of product quality. Indeed, workers or staff employees who complain about a city's services can be called "whistle blowers" and have even lost their jobs. City governments must listen to their city workers who have identified process problems that need correcting and to the ideas they have for correcting them. City governments should embrace JIT principles that will require self-correction of worker-generated defects and seek to maintain a 100 percent quality inspection of the services a government provides to citizens. Because of the size and diverse nature of city facilities, the JIT principle of requiring routine maintenance and housecleaning duties should be delegated to workers and made into a habit by management reinforcement. All city governments should seek continual quality improvement in the services they offer, and as they are assured a long life, they should seek a long-term commitment to quality control efforts performed for their citizens.

7. Service System Output Cannot be Mass Produced   An advertising agency developing an ad campaign for a company is an example of a service system (the ad agency) providing intangible service products (advertising research and development). An advertising agency researches and develops ideas for the efficient use of promotion methods like television ads, printed ads, radio ads, in-store advertising materials, etc. that can help sell a client's products. The information product they deliver to their customers is unique and specific for that client. The JIT inventory, production, and quality principles that can be applied are the same as those previously stated for the management consultant operation. JIT quality principles are of particular importance for an advertising agency. This type of operation generates an information product that is almost always easier to measure than the management consultant information product by fairly rapid and objective measures of the market response of the client's product sales, if increasing sales is the objective of the advertising efforts. An advertising agency's service products also consist of a group effort of many different types of professions (i.e., market researchers, advertising experts, product development experts, etc.) more so than general management consultants who may work as individuals. Building this necessary group effort toward improving quality product is encouraged by JIT principles like making quality everybody's responsibility to ensure everybody's contribution, seeking a high level of visual management on quality to reinforce the quality attitude, maintaining strict product quality control compliance to ensure group members are reminded of the agency's dedication to quality, giving the group members the authority to share in the control of product quality to show the agency's trust in their judgment, requiring self-correction of group member-generated defects to locate problems in process quality for solution process improvement, and maintaining a 100 percent quality inspection of products to ensure product quality for the customer and find opportunities for improvement of the company.

8. Service System Workers Exercise a High Degree of Personal Judgment in Performing Their Work Assignment   A customer service department agent handling customer complaints is an example of a service system (the agent) providing an intangible service product (customer service). The experience and knowledge agents bring to their job in handling complaints is a major ingredient of a service product provided to a customer. Organizations that provide this service (which are growing in number in U.S.) can also employ JIT principles in a variety of ways. Customer problem-solving information can be organized by management in an orderly manner, by categorizing a list of solutions for specific, recurring problems. This will provide agents, who will more easily identify a customer problem as falling into a recognized category, with a predefined suggested solution for that particular problem. On the other hand, this organized information does not have to limit the individual agent's JIT improvement contribution, but only help to add some consistency (a sign of product quality) in dealing with a more commonly occurring set of problems confronted in this system. (We will be discussing the use of "expert system" computer technology that can help organize and deliver information quickly as a means of implementing the JIT principles in the next section.) Another important JIT principle in customer service jobs often ignored by management is allowing workers to determine production flow. Some customer service-based companies actually place time periods in which customer service agents can handle a customer's complaint. Handling complaints is not the type of service product that should be under a quota or limited time frame. Instead, management should, and sometimes does, look upon the interface with a customer as an opportunity to complete their service product or add the extra quality that was left out in the plant or office. Many companies collect very useful quality control information on defects during customer complaint sessions. The JIT principle of allowing workers to determine production flow is a good argument for allowing customer service agents to spend as much time as they feel necessary to handle a complaint and provide a high level of customer service quality. Since the information that customer service agents collect is countable and has an impact on production processes and products, the JIT quality principle of maintaining process control can be implemented to record work contributions and motivate further quality improvements by agents. The impact of information can also be kept track of in a project-by-project record to reinforce the JIT principle of seeking a long-term commitment to quality control efforts.

9. Service Systems are Labor Intensive   A teacher giving a lecture is an example of a service system (the teacher) providing an intangible service product (education). The teacher's job is a very labor-intensive service product. While JIT principles can be applied in this type of operation, little of the JIT inventory management principles applicable to this type of service system and characteristic, except where they relate to the goods of books and materials, are used in the education.

Of the JIT production management principles, some of the most applicable to this type of service system and characteristic include: seeking a synchronized pull system, using automation where practical, seeking improved flexibility in workers, allowing workers to determine production flow, improving communication and visual control, and cutting production lot size. Most of these JIT principles typify grade schools in the U.S. Grade schools hire teachers in accordance with demands of the student population (synchronized pull), use automation, like television education systems, to augment classroom education, seek improved flexibility in teachers' knowledge by rewarding those who go back to college for additional education, allow teachers to decide if students need additional training (remedial education, not promoting students, etc.), and structure their buildings to permit and facilitate large, massive flows of students (i.e., WIP). One JIT principle grade schools have known from research is true for many decades is that of cutting production lot size. This principle applies to the concept of reduced class size, resulting in an improved service product to students.

All of the JIT quality management principles are applicable to this type of service system and characteristic. JIT school teachers, more than anybody else, use and maintain forms of process control (standardized tests) to monitor their service product of education. School teachers also know quality is everybody's responsibility as teachers commonly visit each others' classes to give suggestions to improve teaching methods; teachers know to maintain strict product quality control compliance by striving to help their students to achieve stated academic performance standards; their jobs require self-correction of worker-generated defects in that they may face the same student who fails their class again for another year; they must maintain a 100 percent quality inspection of products from their daily activities with students; they are required to perform routine maintenance and housecleaning duties to keep their classrooms clean (they also have the student-customer help with this JIT principle sometimes); they should seek continual quality improvement by learning to do a better job each year; their careers require a long-term commitment to quality control efforts. Their principals also can and often do help by seeking a high level of visual management on quality through reward systems and by giving teachers the authority to share in the control of product quality.

10. Service System Facilities are Decentralized and Located Near Customers   A telephone company providing telephone service is an example of a service system (the company) providing an intangible service product (communication). In providing the communications service, the phone company also provides a tangible and decentralized set of tangible goods (phones, wire, etc.). JIT principles can be applied in a variety of ways to this type of operation. All of the JIT principles are applicable to the manufacturing of "goods" that this type of service system provides. Little of the inventory principles are applicable to the intangible communication service product.

Of the JIT production management principles for the intangible service product, some of the most applicable to this type of service system and characteristic include seeking uniform daily production scheduling, using automation where practical, and seeking improved flexibility in workers. JIT phone companies today try to create a daily, uniform production schedule by offering customer discounts to increase demand in low-use periods of the day. They also use automation extensively to improve the product quality and reliability. Phone companies in the U.S. have, during the 1980s, moved from a very large and specialized operation to many decentralized operations. This change occurred as phone companies have moved from a monopolistic situation to a more competitive market. The more specialized nature of personnel in the past, centralized operations is giving way to diverse skilled and market-oriented professionals, embracing the JIT principle of hiring more flexible workers.

Almost all of the JIT quality management principles are applicable to this type of service system and characteristic. Since service can be measured by a variety of statistics (messages completed, customer complaints, etc.), phone companies can maintain process control and make quality everybody's responsibility by seeking a high level of visual management on quality. The management of phone companies can establish quality standards and seek to maintain strict product quality control compliance by giving workers the authority to share in the control of product quality. In situations where service does not meet company standards, management can require self-correction of worker-generated defects, and set programs in place that will seek continual quality improvement and seek a long-term commitment to quality control efforts.

11. Service System Measures of Effectiveness are Subjective A lawn care company is an example of a service system providing intangible service products (taking care of a customer's lawn, flowers, trees, etc.). Measuring the effectiveness or level of service of lawn care companies is very subjective. If a tree dies in their care can a customer really blame the company, or just accept it as "nature's way"? A customer's perception that a lawn care company can make plants grow to miraculous sizes also adds to the subjective evaluation of effectiveness by customers. Despite this subjectivity, JIT principles can be applied in a variety of ways to this type of operation. All of the JIT management principles are applicable to the management of the goods (fertilizer, lawn chemicals, etc.) this type of service system uses to deliver their products. Little of the JIT inventory principles are directly related to the service product of lawn care.

Of the JIT production management principles, some of the most applicable to this type of service system and characteristic include: seeking production scheduling flexibility, seeking a synchronized pull system, using automation where practical, seeking improved flexibility in workers, cutting setup costs, and allowing workers to determine production flow. Workers have a wide range of skills to perform the maximum number of different jobs. This will help increase job flexibility, reduce the number of total people required by reducing idleness caused by an inappropriate balance in worker skills, and save the hire/fire waste of activities for a company as the lawn care jobs change over the season. Many lawn care jobs are known and definable in labor time, and personnel skill requirements, like weekly grass cutting jobs. Within the portion of the total jobs that are well known the company can seek to synchronize their work and labor requirements to avoid wasted resources. JIT lawn care companies should also balance the time saving and quality-enhancing abilities of automated equipment over the flexibility of employing humans. Maintaining a large number of employees not only can be used by the management as a marketing ploy of a personal touch, but also, as JIT principles advise, may provide companies with a source of ideas for improving setup time and costs.

All of the JIT quality management principles are applicable to this type of service system and characteristic. Process control charts can be used to maintain process control on flaws found on customer jobs. Management can seek to make quality everybody's responsibility by establishing systems that reward workers for customer compliments, such as the use of mailable customer comment cards, left after each lawn care job, that can be mailed directly back to management. When poor performance is reported, management should require self-correction of worker-generated defects to reinforce the company's commitment to quality, and seek a long-term commitment to quality control efforts from all employees. Management can also give the workers authority to share in the control of product quality by setting aside time regularly to discuss any problems that may impact on the quality of a job. They may uncover problems with equipment that can be dealt with by implementing the JIT principle of requiring workers to perform routine maintenance and cleaning duties. Management may also uncover improvement in production processes and products from the workers' suggestions that will help to seek continual quality improvement in the company's service products.

12. Service System Quality Control is Primarily Limited to Process Control If a bank were manufacturing a tangible product, it could use a quality control process called acceptance sampling to aid in screening defects from its product. Unfortunately, the bank's processing of a customer's check, while being an example of a service system (the bank) providing an intangible service product (convenience of secure bill paying), is limited to chiefly employing **process control** measures for quality control. There is no tangible product in banks for acceptance sampling, so these types of operations cannot use any sampling methods to prescreen the service product they offer customers. What this means is that bank service products are limited to using the same type of quality control system that is used in all JIT operations. Fortunately, JIT principles can be applied in a variety of ways to this type of operation. Of the JIT inventory management principles, one of the most applicable to this type of service system and characteristic is improving material handling. Methods for moving and improving the flow of customer bank transactions through a bank should always be improved. The acquisition of **automatic teller machines (ATMs)** to better reach customers in stores and locations throughout cities is one way banks are improving transaction flow and improving the service product quality under the JIT quality principle of using automation where practical.

Of the JIT production management principles, some of the most applicable to this type of service system and characteristic include seeking uniform daily production scheduling, seeking production scheduling flexibility, seeking a synchronized pull system, seeking a focused factory, seeking improved flexibility in workers, and allowing workers to determine production flow. Much of the in-house work performed in a bank in processing involves activities that can be planned, given long lead times (e.g. processing banking loans, paper work on customer accounts, preparing financial statements, etc.). By bringing together groups of highly flexible bank officers (skilled and fully authorized to perform a variety of financial processing activities) to work in a type of cell layout, the application of JIT scheduling, focused factory, worker flexibility, and production flow principles can and have saved wasted routing, scheduling, labor, and order processing time [4].

Most of the JIT quality management principles are applicable to this type of service system and characteristic. Banks often chart and use objective measures, such as total customer accounts or total assets, to help monitor and maintain process control; they also should remind all workers, particularly the first line employees like tellers, that quality is everybody's responsibility in the intangible service product of the banking business. At the same time, bank management should give the workers authority to share in the control of product quality by encouraging them to offer suggestions to improve service and at the same time enforcing quality standards by requiring self-correction of worker-generated defects when errors in customer transactions are found in computer audits. Considering the possible importance of any one bank transaction, banks should, and many do, maintain a 100 percent quality inspection of products by performing computer audits at the end of each day to catch teller errors. Indeed, it is interesting to note that JIT principles suggest that not being able to use acceptance sampling will be beneficial and will help to keep an organization achieving the JIT 100 percent quality inspection goal. They should also seek continual quality improvement by finding the causes of defects and preventing them from happening in the future. Finally, banks, which specialize in long-term operations, will be working in their own self-interest by seeking a long-term commitment to quality control efforts that will, as in the case with government operations, help ensure long-term survival.

In summary, the application and integration of JIT principles to service operations (those possessing the twelve service system criteria described above), provide past, present, and future opportunities for improving service operations. As we have discussed, a limited number of JIT principles have been used in service operations for decades. The just-in-time nature of service product production processes and the demand pull nature of the customer market for these service products has naturally created environments where only JIT operations could survive. Now that JIT principles have evolved to broaden application in service operations, managers can use the examples in this section and look for new opportunities to further integrate JIT into their own unique service systems.

In the 1990s there is a greater need for higher quality and greater added value in service products than in the 1980s. To match this need, technologies that embrace the JIT principles will be integrated into service systems. One technology that holds great promise in supporting the information-based service systems is expert systems.

TABLE 7-1 • *Comparison of Artificial and Human Expertise*

(a) Advantages of Artificial Expertise over Human Expertise

Artificial Expertise	Human Expertise
Easy to transfer	Difficult to transfer
Easy to define boundaries	Difficult to define boundaries
Fast to access	Slow to access
Consistent	Inconsistent
Affordable	Expensive
Permanent	Perishable

(b) Advantages of Human Expertise over Artificial Expertise

Artificial Expertise	Human Expertise
Less adaptive	Adaptive
Uncreative	Creative
Inflexible	Flexible
Technical knowledge	Common sense knowledge
Narrow knowledge domain	Broad knowledge domain
Interface input limited	Interface input almost unlimited

A Technology for Implementing JIT Principles in Service Systems

An **expert system (ES)** is a computer software system that attempts to capture the experience, decision rules, and thought processes of experts in a given area of knowledge. Expert systems are developed by securing information to form what is called a **knowledge base** from experts (**domain experts** in a specific field (the **knowledge domain**). Included in the information taken from the domain expert is an understanding of how his or her mind reasons. To model this reasoning process, expert systems often use another type of computer software called **artificial intelligence (AI).** AI is used by the computer to simulate and mimic the thought processes of the human mind. AI software is usually composed of modeling systems (like decision trees from statistics) or if-then logic statements to enable a computer system to reach a conclusion through a process of elimination as experts do. AI system expertise does have some important advantages and disadvantages when compared to human expertise, as listed in Table 7-1. The ES and AI are usually incorporated into and are considered a part of an organization's **management information system (MIS)** as depicted in Figure 7-1. Users, workers, or service agents of the organization can access the ES through computer terminals linked to the MIS.

The combination of ES and AI creates a far more powerful source of information than simple computer-based educational reference or tutorial systems that provide users with education on a subject. For example, suppose a person has a problem with a fuel system of an automobile. The person can use a library computer-based reference system to find books that are related to automotive repair. To obtain the information, the person has to enter subject information into the reference system to locate the books that contain related information. Once the person has actually found the books in the library, he or she must then find the chapters in the books on fuel systems, and finally read through the chapters to find solutions to problems that come closest to the symptoms of the problem exhibited by the car in question. This approach could take days. Instead, the customer could access an ES whose knowledge domain is on automotive repairproblems, and through a question and answer session of perhaps minutes, be given specific strategies for solving the specific fuel system problem.

FIGURE 7-1 • *Integrating Expert Systems and Artificial Intelligence to Support Service System Information Activities*

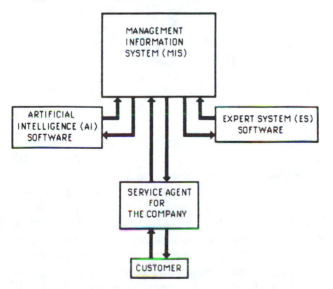

The ES tries to function the same way that any expert's mind functions. Through a process of elimination, the question and answers guide a user logically to narrow the range of possible problems in the knowledge domain down to a specific problem. Once the problem is located, the ES provides possible solution strategies that experts have suggested will solve that particular problem. The solution strategy or strategies are specifically designed to solve the narrowly focused problem identified through the user's interactive session of questions and answers based on the problem symptoms. If more than one solution strategy is offered, it is often accompanied by a statistical measure of reliability that, based on expert judgment, shows individual past success with that strategy. The more advanced ES actually allows these reliability measures, and therefore the solutions offered by the system, to change as more experience with the strategies is collected over a period of time. This advanced feature of ES simulates the growth process of learning from experience.

Expert systems are not without limitations. They are expensive and time consuming to develop. Expert systems are domain specific, they lack the flexibility of humans in dealing with a broad range of problems, and don't offer creativity or original process improvement of ideas beyond their "knowledge domain." Moreover, to apply this technology requires the development of a system limited to areas where genuine experts exist, problems can be defined discretely, experts can articulate their methods, and problems do not require common sense but rather cognitive skills.

Despite these limitations, expert systems have been successfully used in improving manufacturing processes in recent years [6, 7]. Expert systems have been applied in part design [8], in process planning [9], in equipment selection [10], in layout design [11], and inventory model planning [12]. One ES based on GT cells is designed specifically to show how expert systems can be integrated into JIT manufacturing operations [13]. Their application in manufacturing during the 1980s is of a more recent origin when compared with past service applications in the areas of financial planning during the 1970s [14]. The foundation of older service-oriented applications of ES with the more recent applications in manufacturing is pointing toward logical support of JIT in service systems.

An ES is an ideal type of technology for service systems whose service product is chiefly information-based. An ES can also help to integrate JIT principles in these information based service systems. An ES saves the waste of a service agent's time by being able to access information more quickly when requested by customers, and thus, makes agents more productive and efficient. An ES adds value to the information service product it helps to deliver. For example, the ES-generated solution is usually more specific to the nature of the problem (and therefore, adds value because it will be more efficient for the customer's use in eventually solving the problem) than a less experienced human may give a customer. In this sense, it acts like JIT to improve material handling of information. An ES also adds value in that it will seek to ensure consistency in information by logically arriving at the same solution given the same problem symptoms. Human experts, on the other hand, may offer one solution one day, then forget it and suggest another solution the next day for the same problem. Also, the ES supports the JIT principle of giving workers the authority to share in the control of product quality, since it permits agents to freely suggest improved solutions beyond those offered by the system. Being computer based, the ES in combination with an MIS can be used by management to keep track of information and subsequent customer satisfaction to provide management with JIT process control information. An ES also supports the JIT principles of allowing workers to determine production flow, since it is in a supportive role to the service agent and doesn't determine the final or completed timing of service efforts.

Integrating JIT in Administrative Systems

Rationale for JIT in Administrative Systems

There are several reasons for applying JIT principles to administrative systems. One reason for their application is the potential for possible improvement in reducing overhead costs. In breaking down the categories of work activity and their contribution to total manufacturing cost overhead, one study reveals that direct administrative services account for at least 20 percent of total overhead costs, and may represent as much as 35 percent of the total cost of a product to a customer [15]. This rather significant portion of total product costs represents a great opportunity for cost reduction under JIT manufacturing principles that are applicable to administrative settings. As we can see in Table 7-2, there are a number of corresponding target areas where wasteful activities in administrative systems might be reduced by JIT manufacturing principles [16].

TABLE 7-2 • *Joint Target Areas for JIT Principle Application in Manufacturing and Administrative Areas*

Manufacturing Area	*Administrative Area*
Inspecting (quality control)	Proofing
Scheduling (production planning)	Routing
Moving (material handling)	Mailing
Storing (inventory)	Filing

Another reason for using JIT principles in administrative systems concerns the idea of **wholism** in organizations. Most business people realize that an organization is a collection of subsystems (accounting subsystem, marketing subsystem, etc.) that comprise the organization as a whole. So in trying to maximize the performance of an organization, we must view it wholistically; each subsystem must work together in a unified behavior to achieve maximum performance from the organization as a whole. If one subsystem, like manufacturing operations, adopts JIT principles that permit increased performance efficiency may cause bottlenecks in other subsystems, like accounting, which must keep up operations' activities. From a wholistic standpoint, it is better for the whole organization to adopt JIT principles so the improved benefits in one subsystem can be balanced by improved benefits in other subsystems. Experts and research on the implementation of JIT in manufacturing organizations have shown that to receive maximum JIT benefits, organizations should adopt JIT throughout the whole organization [17, 18, 19, 20 and 21].

While the call for the application of JIT in administrative systems was presented in the early 1980s [22], only recently have specific applications in the administrative area started appearing in the literature [16, 23, 24, and 25]. JIT improved efficiency by reducing order filling times, reducing lead times and setup times on all types of administrative activities, reducing backorder logs, reducing personnel, providing faster information flow, and providing higher product quality in administrative services. In effect, by reducing time, effort, and resources used in administrative activities, an organization reduces the non-value-added activities, and improves its organizational efficiency. This improvement in efficiency has for some organizations started the productivity cycling process (discussed in Chapter 1) and helped permit them to more successfully compete in a variety of markets.

Regardless of the rationale chosen to justify JIT principles in administrative systems, their benefits for administrative activities are a proven fact. JIT principles work well in administrative environments for those organizations that know how to apply them. We will discuss how to take advantage of these benefits by integrating JIT principles into administrative systems in the next section.

Implementation Strategies for Implementing JIT in Administrative Systems

Many of the JIT management principles and implementation strategies discussed in Part I of this book can be applied in administrative systems. Administrative systems have inventory (orders that need filling), highly repetitive production processes (order processing procedures), and product flow (the flow of customer orders through departments) like those of manufacturing systems. In manufacturing systems we use JIT principles to reduce inventory and improve production processes. This causes changes in manufacturing operations, which can result in improved product flow by reducing complexity and reducing waste. We seek to accomplish the same goals of simplicity and reduced waste in administrative systems. In this section we will discuss some of the JIT manufacturing implementation strategies that can be used as principles in structuring administrative systems.

To begin the implementation of JIT principles in administrative areas, users need to be familiar with JIT principles. The same JIT manufacturing principles that are presented to the manufacturing personnel should be presented to office and administrative personnel as well. The JIT principles of improved flexibility in workers, improved communication, and making quality everybody's responsibility all support the idea that personnel should share information pertinent to the wholistic operation. It is particularly true that JIT management seeks to use the intelligence of workers to apply JIT principles wherever they can be applied in an operation. It is, therefore, reasonable to assume that administrative personnel might be able to come up with applications unique to their office operations. Such worker-originated ideas for administrative applications of JIT principles have been observed in organizations experimenting with JIT [23, 24].

Once the basic JIT principles are introduced to administrative personnel, management must begin their implementation with specific action-oriented strategies and tactics designed for administrative systems. While there may be an infinite number of specialized strategies that can be uniquely applied to an organization, there are several JIT implementation strategies that can be applied to almost all administrative organizations.

1. Worker or Department Responsibility for Quality Control Workers, or if not feasible, whole departments should be held responsible for their service product's quality. All of the JIT quality management principles can be applied to the individual workers' output, or at a some level of grouping of workers (not more than departmental level) where the processing of administrative work can be identified, and quality inspected and measured. Within departments, the processing of forms or orders often has to go to several work stations where administrative processing is performed. Workers should be required to perform quality control checks (and recording tasks) on prior work station quality efforts to ensure a 100 percent inspection of service products. Much of administrative services involves electronic communication. Automation is making it tactically possible to reduce the quality control inspection effort by using computer technology, like spelling checkers for proofreading. Other sources of quality control information can come from outside the organization. Government agencies are all too happy to provide information on errors in accounting or financial reports. Customers also can be an important source of quality control information, both in measuring quality and finding problem areas. Customer complaints about administrative services, like long lead times in processing orders, lack of personal sales and service calls on customers, or misfiled orders, can all be counted and used as measures for judging the quality of administrative services. They can also be used as statistics to implement process control charting as a quality control tactic. It should be remembered that the more individual these control charts can be (i.e., focused on a worker rather than a group of workers in a department), the more effective they will be in motivating workers to seek continual improvement in their own administrative activities. When poor quality is observed in the delivery of an administrative service that can be corrected, the individual worker or department that is responsible should be made to implement the correction. Everybody in an administrative department should serve as a customer service agent since quality is everybody's responsibility in a JIT administration. All workers should also be asked to perform simple housecleaning activities to keep their work station properly organized and clear of materials (such as work in progress, equipment, personal items, etc.) that waste time by adding confusion to the performance of work activities. Administrative managers must also constantly reinforce quality as a JIT practice through weekly or monthly meetings with workers. When problems surface at these meetings, managers might want to use the tactic of forming a committee or quality control circle to motivate workers to suggest solutions for problems. As administrative workers see that management is making a continual and long-term commitment to quality improvement, they will respond as manufacturing workers have with a continual and valuable source of ideas for product improvement.

   2. Scheduling Administrative Work at Less Than Full Capacity If administrative workers are going to be asked to perform extra quality control, housecleaning, and communication activities each week, then they must be allowed extra time to perform those extra activities. Workers have to know they have the authority to stop their work and bring administrative problems they observe to management's attention. Scheduling at less than full capacity reduces the work load required, permits workers to perform their extra JIT activities, and shows management's commitment to implementing JIT principles. Scheduling at less than full capacity shows management's willingness to sacrifice production for improved quality.

   3. Increase Worker Flexibility All employees should be multifunctional to handle a variety of administrative activities. Research has shown that work specialization in offices can lead to development of behavioral relationships that can constrain work flow [26, p. 10]. For example, if one worker learns from experience the work task preferences of a friend in an office, he or she will be inclined not to give the friend undesirable work tasks. The idea of improved productivity through work specialization can indeed become very costly to an organization that can only operate if everyone comes to work on the same day. JIT administrative managers must increase worker flexibility by increasing the skills and functions of workers. JIT managers can improve administrative assignment flexibility by cross-training personnel and by hiring those workers who have a wide range of office skills. Cross-training also helps to promote the wholistic concept by letting workers understand more of the operation. Cross-training also reinforces the idea of greater job security for workers because of the greater value in being able to offer a variety of skills to the organization. Improvements in technology also force both workers and managers to have a greater range of office skills. It is not uncommon in organizations in the U.S. to have office managers required to know how to type for some jobs that require use of computer-based communication systems.

4. Restructure the Administrative Facility Layout to Simplify Work Methods and to Improve the Application of JIT Principles Classic departmentalized layout structures that separate operations into functional areas should be revised to allow for the application of JIT principles. The application of a GT cell work layout should replace the functional departmentalized layouts whose walls inhibit JIT principles [27]. Walls and departmental barriers cause greater routing, filing, mailing, and proofing. These walls and the work methods that are necessary to support them work against JIT principles. By structuring departments into various groups of workers or GT cell work stations with multifunctional skills, the routing, mailing, and filing activities can be substantially reduced. The amount of processing time can also be substantially reduced and the efficiency of the administrative system improved. As we can see in the example in Figure 7-2, a departmental layout requires a substantial amount of order routing in and out of offices and departments to complete required processing activities. Suppose, instead, we restructure the layout into the three GT cells as presented in Figure 7-3. There would be no departmental walls, and personnel would be cross-trained within departments to be able to perform the multifunctional activities of all three workers in Figure 7-2. The actual routing of an order would only need to be into the area of a single GT cell. All necessary paper work transactions would be performed within the cell, and the order would not leave without being finished. The examples in Figures 7-2 and 7-3 also illustrate the JIT benefit of flexibility that can be achieved by using cell-type layout. Orders in Figure 7-2 only have one channel in which to flow. If one department is backlogged, other offices will become idle and inefficiencies can occur. In Figure 7-3, the administration has the flexibility to allow orders to enter the service system and begin processing in any of the three GT cells. If bottlenecks occur in any one cell, the layout affords users the flexibility to reroute overflow work between cells to better balance work load with available capacity, maximizing operation efficiency. The smaller space and integrated worker environment of a GT cell operation also improves communications by motivating workers to use face-to-face communication rather than written

FIGURE 7-2 • *Order Routing in a Departmental Layout*

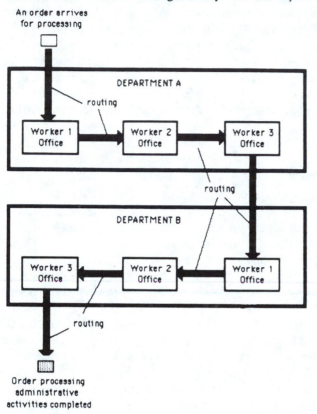

communications requiring filing and processing labor. The GT cell layout also reduces office facility costs in much the same way that reducing inventory in manufacturing reduces facility size. JIT offices do not have to be as physically large, because WIP spends less time in the facility. This reduces the need for filing and filing cabinets. Indeed, the elimination of departments decreases much duplication and can substantially reduce the amount of commonly used equipment, filing cabinets, and storage areas.

5. Increase Standardization of Product Processing Standardizing work activities can be used in the same way that JIT seeks simplicity. Setting up standardized work procedures can save time in cross-training and improve operation efficiency by removing job complexity. Standardizing order forms helps to streamline order taking, reduces order form inventories, and reduces the purchasing of order forms. Standardizing routing systems, where it will not take away from flexibility, removes complexity by transferring WIP through

FIGURE 7-3 • *Order Routing in a JIT GT Cell Layout*

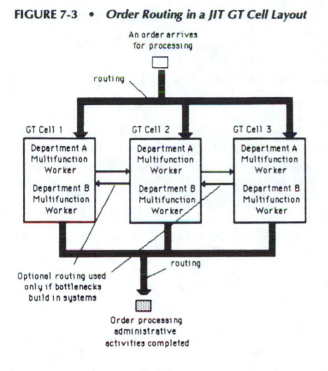

administrative processing channels, and helps the worker to understand and suggest improvements in WIP flow activities.

These JIT strategies and tactics are introductory in nature, but have general applicability to many administrative organizations. While experts suggest that a true JIT application requires implementing considerably more JIT principles than those suggested above, they collectively represent a starting place for JIT application in administrative systems [28]. They address the "time-based competition" demands for administrative tasks and how an organization can use JIT to prepare to meet those demands.

Problems in administrative activities often involve changing long-standing policies, procedures, and work rules. Making changes in policies, procedures, or rules that have been used for many years in an administrative area of an organization is not as easy as changing materials or machines to solve a problem in the manufacturing area. Moreover, a change in one administrative policy can impact others, causing additional problems. JIT administrative management and workers need methods to identify and systematically allocate their limited resources to solve multiple administrative problems. In the next section we will present two JIT methods that can be used to perform problem identification and resource allocation activities.

Methodologies to Aid in Implementing JIT in Administrative Systems

JIT principles seek to invite the uncovering of problems in administrative product quality and production processes. JIT managers need methods to help in efficiently dealing with these problems. To help JIT managers and workers efficiently identify causes of problems, and then prioritize them for problem solving efforts, two methodologies are presented as follows in this section.

The Cause and Effect Diagram Method for Identifying Problems

The **cause and effect diagram** is a graphic aid that can be used to help JIT managers and workers efficiently identify the causes of problems in administrative settings. The idea of the cause and effect diagram is to allow JIT employees to be able to trace known symptoms or effects back to their causes. A sample outline of what a cause and effect diagram looks like is presented in Figure 7-4. Users start with the list of problem symptoms or effects, listing and finding the problem symptoms that best describe the problem they are experiencing. They then follow the arrows in the diagram back to each of the people, equipment, and policy or procedure causes that are related to the symptoms. These diagrams act like pictorial expert systems to help locate the sources of problems by the observed effects.

Obviously, the list of problems has to be work station focused so it doesn't get so large that users are inundated by the information. These diagrams are best used in dealing with the more frequently recurring problems that workers and managers face every day. The diagrams can be placed throughout an office or in strategic locations where problems frequently are encountered. They help both workers and managers to quickly identify possible problems, saving time that can be devoted to problem solving efforts. Consistent with the JIT principle of improving visual control, JIT administrative managers can use the cause and effect diagrams with step-by-step instructions for solving problems that have been encountered during the past. As an historical, problem experience reference point, they also help to improve efficiency in narrowing possibilities when dealing with new problems.

FIGURE 7-4 • *Cause and Effect Diagram*

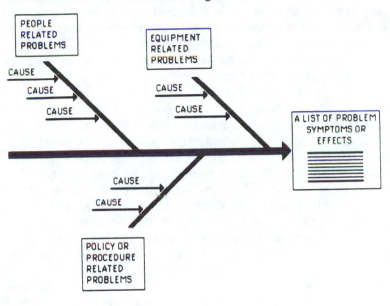

The Pareto Principle Method for Prioritizing Problems

Problem-solving effort is an administrative management resource that should be allocated as efficiently as possible. This resource is usually limited and may not be able to satisfy all of the needs placed upon it. As such, JIT administrative managers need to be able to allocate their limited resources as efficiently as possible to deal with the many opportunities for system improvement motivated by JIT principles of improvement.

One method to help management prioritize JIT administration of problems so they can be resolved in the most efficient manner is based upon the **pareto principle.** The pareto principle simply means that management should allocate most of their resources to solving the most serious problems. This principle seeks to efficiently focus problem-solving resources on the vital few problems that occur most frequently, while leaving the trivial many little problems that occur very infrequently to be dealt with later, after the more serious problems have been resolved. This operates on the same basic idea that makes the ABC inventory classification system efficiently allocate inventory resources in manufacturing operations.

In an administrative setting, managers using JIT principles will find multiple problems need to be solved. Some problems will occur with greater frequency than others. Those problems that occur more frequently should be given a higher priority for solution Those problems that occur less frequently should be given a lower priority. A simple tally of the number of times a problem is reported by workers or customers can be used to establish the priorities. Those problems receiving a greater number of tallies should be given a high priority. Their frequent recurrence reflects the potential for incurring costs to an organization, and by establishing a high priority for these problems, available resources will be correctly channeled to aid in their resolution.

The pareto principle ranking can be modified to consider the frequency weighted cost of problems in cases where the cost of a problem can be accurately assessed. For example, suppose the weekly frequency of occurrence of three problems (Problems A, B, and C) are 4, 2, and 1, respectively. Based solely on the pareto principle of frequency, the rankings of these problems are where A is ranked first, B is ranked second, and C is ranked third. Now suppose the cost of these problems is fairly certain to be $100 for each time A occurs, $150 for each occurrence of B, but when C occurs it costs $1,000. The relatively high expense of Problem C should be considered in ranking its resolution, since its total cost per week is greater to the organization than all of the occurrences of either of the other two problems. To allow for this frequency weighted cost of the problems to be considered in the analysis, all that needs to be done is to multiply the observed frequency of the problem by its assessed cost, then base the rankings on total costs. In this example the total expected weekly cost of Problem A is $400 (4 occurrences x $100), Problem B is $300 (2 occurrences x $150), and Problem C is $1,000 (1 occurrence x $1,000). Based on the frequency weighted total cost for the week, the new rankings of the problems are where C is ranked first at $1,000, A is ranked second at $400, and B is ranked third at $300.

Summary

Service systems in the U.S. are a dominant and growing collection of industries. JIT can play a major role in making service systems operate more efficiently, regardless of how much of the service product is "goods" or "services." We have discussed in this chapter the application of JIT principles to systems that share the various criteria used to characterize service systems. Prior and recent applications of JIT principles in service systems were presented along with suggestions on new areas of possible JIT application. A discussion on expert systems and how they can be used to support and enhance the application of JIT principles in information based service systems.

The administrative activities required in all business operations contribute greatly to the cost of the products that these operations sell to their customers. Administrative activities, while necessarily having to be performed, are viewed by JIT management principles as wasteful. As we discussed in this chapter, the integration of JIT principles to administrative areas can and does cause reductions in required administrative activities. The integration of JIT principles means changing administrative systems throughout an organization to become JIT administrative systems. JIT administrative systems require all workers to be responsible for their own work, thus ensuring greater product quality. JIT administrative systems require the restructuring of facility layouts to ease the application of many of the JIT production principles. Workers must also undergo a change in JIT systems to be more skilled and offer management greater flexibility in their work assignments. JIT system managers must be willing to schedule work at less than full capacity to allow workers to perform JIT activities and seek to standardize work activities to increase worker efficiency. As these changes are implemented and JIT principles become an integrated part of administrative activities of an organization, the JIT principles will motivate workers to seek out new problems that inhibit efficiency in production processes. To help both managers and workers identify problems and then prioritize them for solution effort, this chapter discussed the use of two methodologies: cause-and-effect diagrams and the pareto principle.

This chapter completes Part II of this book. In Part III a JIT simulation game is presented. This game is designed to provide users with some practical experience on the application of JIT principles.

Important Terminology

Acceptance sampling a quality control process used to screen incoming or outgoing products for defects. It is based on a statistical sampling theory that sampling small lots of inventory will reveal information about the entire population.

Administrative services tasks that are performed by workers or managers that support the delivery of an organization's service or manufactured products, but are not directly related to the production of a product.

Administrative system a set of procedures, guidelines, rules, or policies that provide the method by which an administrative service product is to be prepared and/or delivered. It is considered a production process for generating administrative services.

Artificial intelligence (AI) a non-human form of intelligence. Computer software that mimics human decision-making capabilities.

Automatic teller machine (ATM) a computer terminal that allows customers to access their bank accounts and perform banking transactions from remote locations.

Cause and effect diagram a graphic aid used to help identify causes of production problems.

Domain expert a person or persons who possess the expert knowledge that is to be mimicked in a computer expert system.

Electronic sensors devices used to monitor production equipment or products. They generate data that are conveyed electronically to computer systems where software is used to determine if adjustments are necessary to equipment or products. They can also be used to help robots position work in process in automated equipment like numerical control machines and in AS/AR systems.

Expert system (ES) a computer software system that attempts to capture the experience, decision rules, and thought processes of experts in a given area of knowledge.

Goods tangible items that are associated with the delivery of intangible service products.

GT cells a type of facility layout design that permits rapid changes in the sequence of production activities.

Knowledge base a data base of electronically stored information.

Knowledge domain a limited body of knowledge about a specific subject or problem.

Laptop computer a small, portable personal microcomputer designed to be used from a lap.

Management information system (MIS) a collection of computer hardware and software used to communicate information within an organization.

Pareto principle a principle that is based on allocating importance to the most frequently occurring activities. It is used to focus priorities on the most frequently occurring production problems.

Process control a management activity of controlling production processes to achieve a desired output.

Pull system a term used in describing the system by which the use of customer orders pulls inventory through a production system.

Real-time an expression used to describe the lack of lag time between an actual behavior and the report of the behavior being performed from a computer system.

Seiri a Japanese term used to represent the proper arrangement of work centers, inventory, and work activities.

Seiton a Japanese term used to represent orderliness of work centers, inventory, and work activities.

Seiso a Japanese term used to represent cleanliness of work centers.

Service a term used to describe an intangible product or activity.

Service system a term used to describe a system or a service operation that performs economic activities producing an intangible product that adds value or creates utility.

Service sector a U.S. Census Bureau classification for industries that provide a predominantly intangible service product. These industries include transportation and utilities, wholesale and retail trade, finance, insurance, real estate, miscellaneous services, and the government.

Time-based competition a term used to describe a strategy of timing the delivery of a product to meet market demand, for competitive advantage.

Wholistic a term used to describe the view of an organization as being composed of multiple but integrated subsystems that must work together for the benefit of the whole organization.

Work stations administrative equivalent of a manufacturing work center where administrative activities are performed by workers.

Discussion Questions

1. What is the difference between goods and services? How is this difference related to service systems in general? Explain.

2. What are the twelve service system criteria? How are they related to service systems in this chapter?

3. How many of the twelve service system criteria might apply to a fast food restaurant like McDonald's or Burger King?

4. Which of the 24 JIT management principles discussed in Chapters 2, 3, and 4 are mentioned as being applicable to the service system of a medical doctor? Which are not mentioned as being applicable? Can you think of a situation where those that are not applicable might actually be applicable? Explain.

5. Which of the 24 JIT management principles discussed in Chapters 2, 3, and 4 are mentioned as being applicable to the service system of a management consultant? Which are not mentioned as being applicable? Can you think of a situation where those that are not applicable might actually be applicable? Explain.

6. Which of the 24 JIT management principles discussed in Chapters 2, 3, and 4 are mentioned as being applicable to the service system of a fresh fruit vendor? Which are not mentioned as being applicable? Can you think of a situation where those that are not applicable might actually be applicable? Explain.

7. Which of the 24 JIT management principles discussed in Chapters 2, 3, and 4 are mentioned as being applicable to the service system of a fast food vendor? Which are not mentioned as being applicable? Can you think of a situation where those that are not applicable might actually be applicable? Explain.

8. Which of the 24 JIT management principles discussed in Chapters 2, 3, and 4 are mentioned as being applicable to the service system of an amusement

park? Which are not mentioned as being applicable? Can you think of a situation where those that are not applicable might actually be applicable? Explain.

9.   Which of the 24 JIT management principles discussed in Chapters 2, 3, and 4 are mentioned as being applicable to the service system of a government office? Which are not mentioned as being applicable? Can you think of a situation where those that were not applicable might actually be applicable? Explain.

10. Which of the 24 JIT management principles discussed in Chapters 2, 3, and 4 are mentioned as being applicable to the service system of an advertising agency representative? Which are not mentioned as being applicable? Can you think of a situation where those that were not applicable might actually be applicable? Explain.

11. Which of the 24 JIT management principles discussed in Chapters 2, 3, and 4 are mentioned as being applicable to the service system of a customer service department? Which are not mentioned as being applicable? Can you think of a situation where those that were not applicable might actually be applicable? Explain.

12. Which of the 24 JIT management principles discussed in Chapters 2, 3, and 4 are mentioned as being applicable to the service system of a grade school? Which are not mentioned as being applicable? Can you think of a situation where those that were not applicable might actually be applicable? Explain.

13. Which of the 24 JIT management principles discussed in Chapters 2, 3, and 4 are mentioned as being applicable to the service system of a telephone company? Which are not mentioned as being applicable? Can you think of a situation where those that were not applicable might actually be applicable? Explain.

14. Which of the 24 JIT management principles discussed in Chapters 2, 3, and 4 are mentioned as being applicable to the service system of a lawn care company? Which are not mentioned as being applicable? Can you think of a situation where those that were not applicable might actually be applicable? Explain.

15. Which of the 24 JIT management principles discussed in Chapters 2, 3, and 4 are mentioned as being applicable to the service system of a bank? Which are not mentioned as being applicable? Can you think of a situation where those that were not applicable might actually be applicable? Explain.

16. How does an expert system support JIT principles? Which ones? How can an expert system support the integration of JIT principles in service systems?

17. What activities in administrative areas are similar to activities in manufacturing? Give examples.

18. What rationales can be used to justify the integration of JIT principles in administrative areas? Explain.

19. How does the application of JIT principles change what an administrative worker does?

20. How does the application of JIT principles change what an administrative manager does?

21. How does the application of JIT principles change an administrative facility?

22. What is a cause and effect diagram? How do we use it in implementing JIT principles?

23. What is the pareto principle? How do we use it in implementing JIT principles?

Problems

Because of the importance of "process control" in service organizations, the questions in this section are focused on their use, even though charting was not presented in this chapter to avoid duplication. (Students will have to read Chapter 4 material on the methodology of "process control" charting to complete these problems.)

1. A telemarketing company offers telemarketing services to a variety of mail order businesses who use televison to promote and sell products. The telemarketing services include personnel who take orders for customer businesses over phones 24 hours a day. Customers who call in orders receive information on their purchases, including the in-stock availability, order processing, and the cost of the item. It is a common occurrence for phone customers to cancel orders during a phone call when they learn of temporary stock-outs, excessive lead times, or excessive costs for items desired. While scripted marketing messages are used by staff in taking orders, some flexibility for staffers to use their own sales skills to help complete customer orders is allowed. For each order taken over the phone, the customer businesses pay a prearranged fee. The more orders taken, the more money the telemarketing company makes. The phone calls coming into the telemarketing center are given to any person who is not currently handling an order in order of the phone system wiring (i.e., the first phone will invariably

be kept busy while later phones might be continually unused). The telemarketing personnel are paid a fixed salary. Suppose the telemarketing company decides to implement a process control charting system to keep track of the number of calls each member of the staff receives each day. The c-chart is chosen for monitoring the process control activities by counting the number of total calls each day. These c's will be plotted each day and posted by the work center as a JIT quality control measure. How is measuring total calls an unfair JIT measure of quality of the company's service? How might measuring the calls satisfactorily completed each day be a more fair JIT measure of quality of the company's service? Explain. What JIT quality principles might apply to this type of service operation? Explain how they apply.

2. A hair stylist running a system of hair care stores wants to implement a process control system. The number of staff at each store ranges by as much as 400 percent. The hair stylist wants a system that would indicate which of the stores has quality problems that warrant monthly attention, and which stores do not need attention. The hair stylist decides to use a p-chart system. To implement this system, each customer will be given a point-of-service comment card that asks the customer to indicate if the service was either adequate or not adequate during a current visit. The cards would then be mailed directly back to the hair stylist. For each store, the cards are collected on a monthly basis, and the "p" proportion of inadequate card evaluations can be plotted on individual charts for each store. These can then be compared based on the past month's performance in each store, and also compared between stores to determine which store the hair stylist should visit for quality control correction purposes. Is this an unfair or fair JIT measure of quality for the store's service? Explain. What JIT quality principles might apply to this type of service operation? Explain which ones and how they apply.

3. A teacher's overall student evaluation score on a continuous five point scale (where 1 is "good" and 6 is "bad") has a career mean score of 2.4, with an average range of 0.5 based on a sample size per class of only 8. The teacher would like to establish a process control system to monitor how well the students perceive the teacher's efforts as a function of the overall student evaluation score. What process control chart should be used to monitor quality in this situation? Explain your choice. What is the quality standard of this system? What are the resulting UCL and LCL for this system? Draw the chart. Suppose the teacher implements some JIT quality management principles. A new sample of students results in the following student evaluations: 2, 2, 2, 1, 1, 3, 1, 1, 3, and 1. What are the revised UCL and LCL? Plot the new sample on the revised process control chart. Has the

adoption of the JIT principles improved the quality of the teacher's performance? Explain your answer.

4. A home delivery service pizza company wants to establish a process control chart system to monitor the promptness of their delivery service. The company prepares cooked pizzas, and delivers them to surrounding residents of a metropolitan city. The company decides to establish $\bar{x}$- and R-charts to monitor the quality of the delivery service. To develop the charts, data on delivery times needed to be collected. One block in the city is used as a test site as it represents a fixed point to determine the variability in delivery times. The idea is that delivery from the pizza company to customers on a specific block should be fairly consistent, since the distance is fixed. To prepare the charts, four days of pizza sales are sampled. For each of the four days, the delivery times in minutes for five deliveries were collected. The data are presented below:

Day Number	Observed Delivery Lead Times in Minutes
1	32, 35, 35, 38, 40
2	32, 29, 32, 34, 33
3	34, 33, 33, 36, 34
4	33, 32, 35, 34, 27

What are the $\bar{x}$- and R-charts' upper and lower control limits for this company's service product? Draw and plot the sample averages and ranges from the data above on each of the two charts. Suppose we implement JIT quality control principles, and allow a period of time to pass. If we take another day's sample of five deliveries, and it turns out that the times for the deliveries in minutes are: 32, 33, 32, 32, and 32, what would you conclude about the JIT implementation? Show your work by plotting the revised, appropriate values on both process control charts. How is this chart really measuring product quality? Explain.

5. A movie theater wants to establish a process control chart system to monitor the quality of the entertainment service product. The theater operates seven days a week from 4:00 p.m. to midnight. They decide to establish a p-chart to monitor quality as measured by a ten-question customer service comment card that is to be filled out by patrons as they leave the theater. The cards are dropped in convenient boxes near theater exits. Each asks patrons to check either "yes—service is good" or "no—service is bad" for each of the ten questions dealing with the quality of the service product the theater offers. The "p's" will be determined from the "no's" of a sample of 8 cards for each

of 10 performances. The data collected for the chart construction are presented below:

Performance No.	Observed Number of "No" Responses from Eight Patrons
1	6, 5, 1, 1, 1, 1, 2, 2
2	2, 1, 7, 0, 3, 3, 3, 2
3	4, 2, 3, 1, 4, 4, 3, 1
4	2, 1, 1, 2, 3, 1, 3, 7
5	4, 5, 1, 1, 0, 1, 4, 5
6	2, 2, 4, 4, 6, 1, 2, 0
7	4, 1, 3, 2, 1, 6, 0, 0
8	1, 1, 0, 0, 3, 1, 0, 0
9	4, 2, 3, 1, 0, 4, 3, 1
10	0, 1, 1, 2, 3, 1, 3, 0

What are the p-chart's upper and lower control limits for this theater's service product? Draw and plot the control limits on a chart for the data above. Suppose we implement JIT quality control principles and allow a month to pass. If we take another performance's sample of eight patron comment cards, and they turn out to be: 2, 3, 1, 1, 0, 0, 1, and 0, what would you conclude about the JIT implementation? Show your work by plotting the revised, appropriate values on the process control chart. How is this chart really measuring product quality? Explain.

6. A company's administrative operation adopts JIT principles. After several months of solving implementation problems, the workers settle on a program of problem identification with order processing activities. Unfortunately, the workers find more problems than can be solved by the manager's available resources. The administrative manager wants to decide which of five problems (Problems A, B, C, D, and E) should be solved first, which second, and so on. The frequency with which these problems reoccurred during one week turned out to be 12 times for A, 34 times for B, 5 times for C, 10 times for D, and 39 times for E. What is the pareto principle derived ranking of these problems? Which problem should be solved first? Why?

7. After adopting JIT principles in an administrative office, workers end up bombarding the office manager with work flow ideas for solving problems. Unfortunately, the workers find more problems than the manager can afford to solve with available resources. The office manager now has to choose

which of eight problems (Problems A, B, C, D, E, F, G, and H) should be solved first, which second, and so on. The frequency with which these problems reoccurred during one week turned out to be 9 times for A, 21 times for B, 23 times for C, 4 times for D, 6 times for E, 8 times for F, 30 times for G, and 2 times for H. What is the pareto principle derived ranking of these problems? Which problem should be solved first? How does the resulting ranking illustrate the basic idea behind the pareto principle in this problem? Explain.

8. A gas station owner finds that each day the same set of problems disrupt the administrative activities of the operation. They include phone calls, parts problems, etc. These problems also are costing the gas station owner valuable time and money. A total of five problems (Problems A, B, C, D, and E) is identified. The daily occurrence of these problems is tallied by the owner. The tallies for the occurrence of the problems are 6 times for A, 3 times for B, 1 time for C, 14 times for D, and 15 times for E. The cost each time a problem occurs is assessed at $10 for A, $2 for B, $1 for C, $12 for D, and $15 for E. What is the pareto principle ranking of these problems if only frequencies are used to establish the ranking? What is the ranking of these problems if the cost weighted frequencies are used to establish the ranking?

9. An office manager finds that each day the same set of ten problems (Problems A, B, C, D, E, F, G, H, I, and J) is disrupting the administrative activities of the office. The weekly occurrence of these ten problems is tallied by the office manager, and is 12 times for A, 25 times for B, 34 times for C, 4 times for D, 15 times for E, 2 times for F, 7 times for G, 8 times for H, 4 times for I, and 23 times for J. The cost each time the problem occurs is assessed at $100 for A, $22 for B, $16 for C, $52 for D, $25 for E, $10 for F, $13 for G, $160 for H, $502 for I, and $65 for J. What is the pareto principle derived ranking of these problems if only frequencies are used to establish the ranking? What is the pareto principle derived ranking of these problems if the cost weighted frequencies are used to establish the ranking?

References

[1] J. B. Dilworth, *Operations Management*, 4th ed., McGraw-Hill, New York, NY, 1992, chapters 1 and 15.
[2] E. S. Buffa and R. K. Sarin, *Modern Production/Operations Management*, 8th ed., John Wiley & Sons, New York, NY, 1987, chapter 16.
[3] R. G. Murdick, B. Render, and R. S. Russell, *Service Operations Management*, Allyn and Bacon, Boston, MA, 1990.
[4] J. Y. Lee, "JIT Works for Services Too," *CMA Magazine*, Vol. 64, No. 6, 1990, pp. 20–23.

[5] R. G. Conant, "JIT in a Mail Order Operation Reduces Processing Time from Four Days to Four Hours," *Industrial Engineering*, Vol. 20, No. 9, 1990, pp. 34–37.

[6] A. J. Taramina, "Expert Systems in Manufacturing," *Production and Inventory Management Review and APICS News*, Vol. 10, No. 12, 1990, pp. 42–45.

[7] J. C. Giarratano, C. Culbert, and R. T. Savely, "The State-of-the-Art for Current and Future Expert Systems Tools," *ISA Transactions*, Vol. 29, No. 1, 1990, pp. 17–25.

[8] G. H. Schaffer, "Artificial Intelligence—A Tool for Smart Manufacturing," *American Machinist and Automated Manufacturing*, August 1986, pp. 84–94.

[9] C. Hayes and P. Wright, "Automating Process Planning: Using Feature Interactions to Guide Search," *Journal of Manufacturing Systems*, Vol. 8, No. 1, 1989, pp. 1–15.

[10] C. J. Malmborg, B. Krishnakumar, G. R. Simons, and M. H. Agee, "EXIT: A PC-Based Expert System for Industrial Truck Selection," *International Journal of Production Research*, Vol. 27, No. 6, 1989, pp. 927–941.

[11] S. S. Heragu and A. Kusiak, "Analysis Systems in Manufacturing Design," *IEEE Transactions on Systems, Man, and Cybernetics*, Vol. SMC-17, No. 6, 1987, pp. 898–912.

[12] M. Parlar, "EXPIM: A Knowledge-Based Expert System for Production/Inventory Modeling," *International Journal of Production Research*, Vol. 27, No. 1, 1989, pp. 101–118.

[13] G. Sheng-Hsien and J. T. Black, "An Expert System for Manufacturing Cell Control," *Computers and Industrial Engineering*, Vol. 17, No. 1–4, 1989, pp. 18–23.

[14] D. A. Waterman, *A Guide to Expert Systems*, Addison-Wesley Publishing Co., Reading, MA, 1986.

[15] J. L. Funk, "A Comparison of Inventory Cost Reduction Strategies in a JIT Manufacturing System," *International Journal of Production Research*, Vol. 27, No. 7, 1989, pp. 1065–1080.

[16] T. Billesbach and M. J. Schniederjans, "Applicability of Just-In-Time Techniques in Administration," *Production and Inventory Management Journal*, Vol. 30, No. 3, 1989, pp. 40–45.

[17] R. J. Schonberger, *Building a Chain of Customers: Linking Business Functions to Create a World Class Company*, The Free Press, New York, NY, 1990.

[18] L. C. Giunipero and W. K. Law, "Organizational Changes and JIT Implementation," *Production and Inventory Management Journal*, Vol. 31, No. 3, 1990, pp. 71–73.

[19] M. M. Helms, "Communication: The Key to JIT Success," *Production and Inventory Management Journal*, Vol. 31, No. 2, 1990, pp. 18–21.

[20] J. Malley and R. Ray, "Informational and Organizational Impacts of Implementing a JIT System," *Production and Inventory Management Journal*, Vol. 29, No. 2, 1988, pp. 66–70.

[21] R. Natarajan and J. D. Weinrauch, "JIT and the Marketing Interface," *Production and Inventory Management Journal*, Vol. 31, No. 3, 1990, pp. 42–46.

[22] R. J. Schonberger, *Japanese Manufacturing Techniques: Nine Hidden Lessons in Simplicity*, The Free Press, New York, NY, 1982.

[23] J. Barrett, "IE's at CalComp are Integrating JIT, TQC and Employee Involvement for 'World Class Manufacturing'," *Industrial Engineering*, Vol. 20, No. 9, 1988, pp. 26–32.

[24] J. Y. Lee, "JIT Works for Services Too," *CMA Magazine*, Vol. 64, No. 6, 1990, pp. 20–23.

[25] R. G. Conant, "JIT in a Mail Order Operation Reduces Processing Time from Four Days to Four Hours," *Industrial Engineering*, Vol. 20, No. 9, 1988, pp. 34–37.

[26] P. A. Strassman, *Information Payoff*, The Free Press, New York, NY, 1985.

[27] J. J. Feather and K. F. Cross, "Workflow Analysis, Just-In-Time Techniques Simplify Administrative Process in Paperwork Operations," *Industrial Engineering*, Vol. 20, No. 1, 1988, pp. 32–40.

[28] P. H. Zipkin, "Does Manufacturing Need a JIT Revolution?" *Harvard Business Review*, January-February, 1991, pp. 40–50.

CASE 7-1

Gassing Up To JIT

The Gasorama Service Station (GSS) has just been purchased by a new owner. GSS is both a full and self service gas station, with 4 gas pumps for full service and 16 pumps for self service. The gas station portion of the business takes up over 20,000 square feet of building space (not including lot space or overhangs protecting gas pumps). The gas station offers customers a full range of automotive mechanical services. In addition to the gas station services, GSS also includes a small grocery store that contains the usual convenience store products. The grocery store takes up about 300 square feet of building space. The entire GSS facility is open 24 hours a day, 7 days a week throughout the year. A single building is used to house both the gas and grocery store. The location of the facility is ideal to attract customers 24 hours a day. It is near an interstate highway off-ramp and is on the busiest street in the city.

GSS employs a total of 17 people distributed in three eight-hour work shifts. Six people work the morning shift from 8:00 a.m. to 4:00 p.m., six people work the evening shift from 4:00 p.m. to midnight, and five people work the late night shift from midnight to 8:00 a.m. Each shift has two highly skilled mechanics, one of whom is responsible for managing the station. Only the skilled mechanics are allowed to do mechanical work on cars. The skilled mechanics are not allowed to pump gas or work in the grocery store. The other workers are unskilled and are only capable of pumping gas or selling food in the grocery. Unskilled workers are assigned jobs at the pumps or in the grocery store at the beginning of the day, and are not allowed to switch later in the day. Prior management felt that the more highly skilled workers who were being paid about 25 percent more per hour should only work in the higher skilled jobs for which they were hired. Also, they felt that unskilled workers should not be allowed to work on cars for fear of losing customers or causing lawsuits for poor work. The workers are paid to do a specific job, nothing more.

Under prior contracts with out-of-state suppliers, three months of "goods" like tires, spark plugs, and oil filters are sent at one time. Quantity discounts on large lot ordering was viewed as a desirable opportunity to lower unit costs. Prior management felt a quick physical audit of inventory every three months was the most efficient way to minimize the paperwork of ordering. When a shortage occurred of a special item, a special rush order was placed with local vendors for that particular individual item. The sometimes excessive cost of these rush items was paid mostly by the customer. In cases involving a good customer, GSS would pay the expediting charges to show interest in keeping that customer coming back. GSS has massive gas storage capacity and receives weekly shipments of gas from its distributor when other stations order on a twice-weekly basis. The occasional complaint from customers about gas lacking octane because it isn't fresh is more than offset by the convenience of not having to place the second order each week. Paper work was avoided by the prior management at all costs to reduce inefficient waste of management's time.

The new owners had been researching the GSS operation for six months prior to the purchase of the operation. They found that at different times of the day, long lines of customers waiting for service form at various points in the operation. In the morning, lines in the grocery store form for people waiting to pay for breakfast cakes, coffee, and donuts, while there is little activity in the mechanics' docks. In late afternoon, commuter cars stack up, waiting for gas service at the pumps, while there is little activity at the grocery store. In the evening, customers pile up at the service docks wanting to obtain mechanical service work. The new owners have on many occasions observed as many as 50 percent of the customers arriving for service never receive it because they leave, frustrated at having to wait too long in a line.

GSS has been closed down for the last two months during sale negotiations. While all of the previous employees found other jobs, the labor market in the area is capable of providing whatever worker talent the new owners want in the operation.

Case Questions

1. What JIT principles can be applied to this service operation? Explain how each could be applied to the "goods" and "services" of GSS.

2. How will the implementation of JIT principles help to improve the customers' service products from GSS? Explain.

CASE 7-2

Serving JIT

The U-Eat'em Grocery Store (UGS) is located in Clayton, Missouri, and offers a variety of food store items to customers. UGS's market over the years has evolved to service the high income and upper economic class of customers. The food items in the store and the focus on customer service are structured to cater to the specific needs of the special clientele.

The basic layout of UGS is divided into four departmental areas of canned goods, fresh fruit and vegetables, meats, and a deli. So specialized is UGS that each department is separately managed and maintained by separate groups of individual workers. For example, separate workers only stock canned goods and are never asked to perform work activities in any of the other departments. Also, each department is autonomous, setting its own reward systems, salaries, working hours, and vacation schedules. The owner of UGS has always felt that the competitive environment of each department seeking profits on its own would maximize profits for the operation as a whole.

The owner of UGS attends a national meeting of grocery store owners and is surprised to find other stores of similar size and sales are making much more profit than UGS. The owner asks a few favorite customers questions about what they dislike about UGS. Invariably, the response is centered on a variety of quality service issues. One customer complains about parts of the store looking dirty while other parts are spotlessly clean. The contrast between departments makes differences all the more unseemly. Another customer complains about the poor orderliness of items the store stocks. The customer couldn't find some items that were previously in the store, and even when the location of the item is found everything is mixed together or appears sloppy.

The owner is quite concerned and decides to take a survey of 25 randomly selected customers who are observed to have shopped in all four departments. These customers are asked to fill out the customer quality service questionnaire in Exhibit 7-1. The customers are asked to subjectively assess each of the departments on a 1 to 5 scale (a "1" means bad service is provided and a "5" means good service). The frequency of responses (the value in the parentheses next to the scoring number) for the 25 customers is presented in Exhibit 7-2.

EXHIBIT 7-1 • *Customer Quality Service Questionnaire*

Canned Goods Department Questions	*Score on Questions (Circle One)* BAD				GOOD
1. If needed, did your server in this department satisfy your customer needs (e.g., timely response to questions, directions to find canned goods, etc.)?	1	2	3	4	5
2. Does this department's appearance meet with your expectations on orderliness (e.g. cans in shelves, cans in right locations, etc.)?	1	2	3	4	5
3. Does this department's appearance meet with your expectations on cleanliness (e.g., opened cans, noticeable dirt, etc.)?	1	2	3	4	5

Fresh Fruit & Vegetables Department Questions

1. If needed, did your server in this department satisfy your customer needs (e.g., timely response to questions, cooking instructions, etc.)?	1	2	3	4	5
2. Does this department's appearance meet with your expectations on orderliness (e.g., fallen fruit on the floor, vegetables in right locations, etc.)?	1	2	3	4	5
3. Does this department's appearance meet with your expectations on cleanliness (e.g., rotting vegetables, noticeable dirt, etc.)?	1	2	3	4	5

Meat Department Questions

1. If needed, did your server in this department satisfy your customer needs (e.g., timely response to questions, cooking instructions, etc.)?	1	2	3	4	5
2. Does this department's appearance meet with your expectations on orderliness (e.g., meat packages orderly in cases, meat and poultry in right locations, etc.)?	1	2	3	4	5
3. Does this department's appearance meet with your expectations on cleanliness (e.g., rotting meats, bloody packaging, bloody butchers, etc.)?	1	2	3	4	5

Continued

EXHIBIT 7-1 *Continued*

	Score on Questions (Circle One)				
Deli Department Questions	*BAD*				*GOOD*
1. If needed, did your server in this department satisfy your customer needs (e.g., timely response to questions, customer service done with a smile, etc.)?	1	2	3	4	5
2. Does this department's appearance meet with your expectations on orderliness (e.g., trays of open food orderly in cases, prices current on trays of food, etc.)?	1	2	3	4	5
3. Does this department's appearance meet with your expectations on cleanliness (e.g., spoilage observed, open food tray contents spilled together, etc.)?	1	2	3	4	5

EXHIBIT 7-2 • *Customer Quality Service Questionnaire Responses* [*]

Canned Goods Department Questions

1.	1	2 (2)	3 (4)	4 (5)	5 (14)
2.	1	2	3 (8)	4 (6)	5 (11)
3.	1	2	3 (5)	4 (4)	5 (16)

Fresh Fruit & Vegetables Department Questions

1.	1 (20)	2 (4)	3 (1)	4	5
2.	1 (8)	2 (9)	3 (3)	4 (5)	5
3.	1 (4)	2 (15)	3 (6)	4	5

Meat Department Questions

1.	1 (4)	2 (8)	3 (8)	4 (4)	5 (1)
2.	1 (10)	2 (4)	3 (3)	4 (2)	5 (6)
3.	1 (9)	2 (8)	3 (2)	4 (3)	5 (3)

[*]Value in "( )" is the frequency of customers who chose the related 1 to 5 score for each of the three questions per department.

Deli Department Questions

1.	1 (3)	2 (1)	3 (4)	4 (5)	5 (12)
2.	1 (4)	2 (2)	3 (3)	4 (1)	5 (15)
3.	1 (2)	2 (4)	3 (5)	4 (8)	5 (6)

**EXHIBIT 7-3 • Customer Quality Service Questionnaire Responses
From Second Survey[*]**

Canned Goods Department Questions

1.	1 (1)	2 (2)	3 (2)	4 (3)	5 (17)
2.	1	2 (2)	3 (2)	4 (4)	5 (17)
3.	1	2 (2)	3 (3)	4 (3)	5 (17)

Fresh Fruit & Vegetables Department Questions

1.	1 (20)	2 (3)	3 (1)	4 (1)	5
2.	1 (10)	2 (6)	3 (3)	4 (3)	5 (3)
3.	1	2 (18)	3 (6)	4 (1)	5

Meat Department Questions

1.	1 (8)	2 (4)	3 (4)	4 (8)	5 (1)
2.	1	2 (4)	3 (13)	4 (2)	5 (6)
3.	1 (9)	2	3 (10)	4 (3)	5 (3)

Deli Department Questions

1.	1	2 (1)	3 (4)	4 (5)	5 (15)
2.	1	2(2)	3 (3)	4 (5)	5 (15)
3.	1 (2)	2	3 (5)	4 (8)	5 (10)

---

[*]Value in "( )" is the frequency of customers who chose the related 1 to 5 score for each of the three questions per department.

Having collected the information, the owner decides that: (1) the departments' reward systems (i.e., the bonuses for profit) are going to be allocated solely on improving customer perceived quality in an individual department's service, and that (2) UGS is going to start a JIT quality management program as a means of implementing improved quality in the operation. The owner is particularly interested in implementing the process control principle as it relates to both the characteristic nature of the service operation of food stores and to JIT management. The owner specifically wants to implement a process quality control system that would use the customer quality service questionnaires as data. In preparation, each department has its own control chart, so the owner, department managers, and workers can all graphically see changes in customer quality perceptions as they take place over time.

A month after the implementation of JIT principles the customer quality service questionnaire is again used on a different set of 25 customers. The data from this survey are presented in Exhibit 7-3.

Case Questions

1. What JIT quality management principles can be applied to this type of service operation? Explain how each of them applies.

2. What control chart(s) should be used to monitor quality in the various departments? Using the data in Exhibit 7-2, prepare all four charts.

3. What are the revised control charts when only the new data from Exhibit 7-3 are used to determine the UCL and LCL values? Has the switch to JIT principles helped this operation? Explain where it has helped.

CASE 7-3

Administrating JIT

The You Risked It (YRI) Insurance Company of Beavertown, Iowa, has a major insurance office facility located in southern Iowa. The YRI office is responsible for processing all insurance claims forms for the entire nationwide YRI insurance organization. The company's sales experienced rapid increases during the last year, and as expected, record levels in claims during recent months. Over 2,500 forms a day are now flowing into the office for processing. Unfortunately, less than 2,500 forms a day are flowing out. The office operations and administrative activities have so backed up over the last two months that customers are threatening legal action if something isn't done to speed up insurance form processing. Under the current set of administrative systems, human resources, and facilities, nothing can be done to handle the load of work YRI is currently experiencing.

YRI management feels that it is time to revise administrative activities to improve the work flow. They decide to embrace JIT administrative principles. In the short term, part-time personnel are brought in from the sales divisions across the country to help fill in during the changeover period and to work down the backlogged claims.

In the meanwhile, YRI managers and workers are given training on JIT principles by visiting JIT experts. With the JIT principles firmly in mind, YRI managers go about changing job responsibilities to require quality control activities from each employee, and hold each employee responsible for correcting poor quality work when reported. They also start worker training programs that offer a variety of new job skills. Typing skills are taught and required of both workers and managers in an effort to reduce dependence on secretarial pools. Emphasis in cross-training is given top priority to permit workers within departments to be able to process all types of customer claim applications. Management is also reexamining the insurance and claim forms used in the processing areas to see if standardization of forms and processing procedures might save administrative time in training and in WIP. Making changes in work methods and workers leads logically to a need for change in the layout of the YRI facility. This potential change represents a big move from the departmental layout of the past, and will be a costly step. To help ensure the successful implementation of this phase of the changeover, YRI hires a team of office layout experts as consultants.

The team of experts agrees with YRI management that a change in the YRI office facilities is necessary to successfully implement and integrate JIT principles in the operations. The consultants feel they should revise the layout of the operation and work methods to bring them in line with "GT cell" layouts. An example of the typical, current departmental layout and insurance form processing routing flow pattern is presented in Exhibit 7-4. As can be seen in Exhibit 7-4, customer claims arrive for processing at the facility and flow through a series of offices and departments to eventually complete the claim process. While the arrangement of the offices and workers in Figure 7-4 is typical, the entire facility

EXHIBIT 7-4 • _Customer Claim Routing and Departmental Layout_

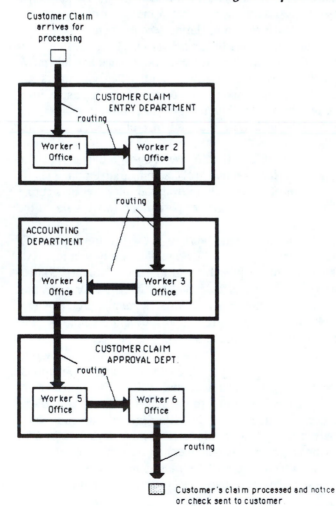

actually operates 20 additional flows, each consisting of the same three departments, but with a different number of office workers performing a variety of different tasks depending on the claim form requirements (larger claims have to be approved by more people). So while all forms must pass through the same three departments (Customer Claim Entry, Accounting, and Customer Claim Approval), the number of workers for each claim may vary from two to four. For this reason, the YRI facility currently maintains a staff of over 180 office workers just to process and route claims.

Case Questions

1. What would the restructured GT cell layout conversion for the typical departmental claim routing layout in Exhibit 7-4 look like? Draw it.

2. What assumptions are you making about personnel and equipment in the new GT cell layout in Question 1 above?

3. How will the new layout from Question 1 permit the integration of JIT principles in the YRI operation? Explain.

4. How will the new layout from Question 1 benefit the YRI operation? Explain.

PART THREE

A JIT Simulation Game

A JIT Simulation Game

Introduction

Every practitioner knows that reading principles and applying them are two different things. This part of the book is designed to help reinforce the learning experience of reading JIT principles through a series of in-class simulations where students actually experience the application of JIT principles.

Students will be asked to perform one or more roles (worker, manager, accountant, material handler, engineer, or inspector) in two different types of manufacturing operations: (1) a JIT operation, and (2) an EOQ operation. The layout and manufacturing rules will differ to represent the nature of the two different types of operations modeled in the simulation. Operating under these two types of operations in separate simulations, students will be asked to produce (assemble) one or two different products. The two simulated products the students will be assembling are slightly different versions of toy houses (let's say representing a company's "product family") and using LEGO™ Brand toy bricks from INTERLOGO A. G. of Switzerland.

The general information on basic production layouts for the JIT and EOQ operations, the basic roles for personnel, and the basic work activities are presented in the next section, "Simulation Game General Information." In the following section, "Experiments in JIT and EOQ Operations," a series of simulation experiments or games are described. Specific changes from the basic layout, production scheduling, and work activities will be described as they relate to the application of different JIT principles. The simulation experiments will always consist of running both a simulated JIT operation and a simulated EOQ operation. During the experiments, data on production rates, time requirements, inventory levels, and product quality will be recorded and compared for the two different types of operations.

Students will be able to enjoy some of the triumphs and failures of manufacturing as they operate under both systems. Most importantly, they will experience some of the real world difficulties in making systems work, and learn how JIT principles can benefit production planning and control activities.

Simulation Game General Information

In running the simulation game the instructor plays the role of "President." The President may change the basic operations described in this section to fit the requirements of the classroom. The President also may dictate production activities, goals, and when games begin and end. Additional rules and guidelines for the basic production activities and player roles for both types of operations are necessary to begin the game.

The Basic JIT Operation

The basic JIT operation (hereafter called the JIT operation) will be used in all of the simulation experiments with little or no modification. The layout of the JIT operation will consist of a straight line flow shop production cell layout of five work centers as presented in Exhibit 1. (It may become a GT cell in some simulations.) Work on either of the two products (Product A or Product B) will begin at Work Center 1. The WIP will go from Work Center 1 to Work Center 2, and so on, until it's finished at the completion of work activities at Work Center 5. From there the finished product will be given to the Quality Control Inspector for inspection. Once approved, the finished product will be given to Material Handler 5 who will start the breakdown of the finished product. After Material Handler 5 performs the necessary work activities, the portion of the finished product is sent to Material Handler 4, then 3, and so on. By the time Material Handler 1 completes the breakdown activities all of the finished product will have been converted back into the various component parts. These component inventory items are given by the material handlers back to their respective work centers. This completes the process of production and inventory replenishment which will permit continuous activity from all participants. The basic roles of each of the six different types of players in the game are as follows:

1. Workers   There are five different workers, one each for the five work centers that make up the JIT operation layout in Exhibit 1. Each worker will be asked to assemble a set of component parts into units of either Product A or Product B. The specific component part assembly work for each work center is presented in Exhibit 2 for Product "A" and Exhibit 3 for Product "B." For example, for Product "A" the worker at Work Center 1 has to assemble five LEGO™ bricks. The color of the brick and the size, represented by number of locking nodes on top of each brick, are indicated in the exhibits.

To facilitate visual production instructions via JIT principles, students may keep the exhibits next to their open books by their work center for reference. Be sure to assemble the right color and size brick, because quality counts. Also consistent with JIT principles, workers are asked to check the quality of the work performed at the prior work center. If material handlers don't do their job, workers should correct them. So, Worker 2 is asked to check Worker 1's assembly job, and Worker 3 should check Worker 2's, and so on. If defects are found by the worker, the item goes back to the previous work center whether

EXHIBIT 1 • *JIT Production Cell Layout and Staff*

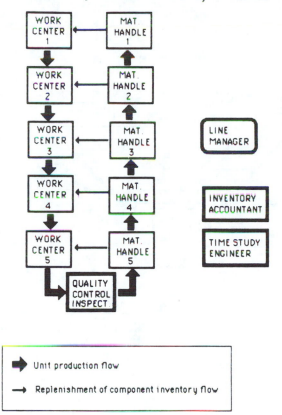

the problem originates there or further back up the line. If defects are found by the Quality Control Inspector, the item is given directly back to the work center where the defect originates so it can be broken down and the defect corrected. This simulates the JIT principle of correcting one's own errors. When the game begins, a total supply of inventory will be sitting in as many trays as the worker has component parts to assemble. Each component part should have its own individual tray and should never be mixed together. When the game begins, workers who have the time can organize the trays or bricks to facilitate their application in production. Workers may also preassemble any of the component parts inventory into their complete subassembly (i.e., the output of the individual work center). For example, Worker 1 can preassemble the five bricks for product A into a subassembly to facilitate WIP. These production improvement activities should be conducted during periods of time when workers are idle because of

EXHIBIT 2 • *LEGO™ Brand Toy House Product "A" Instructions for JIT Cell Layout*

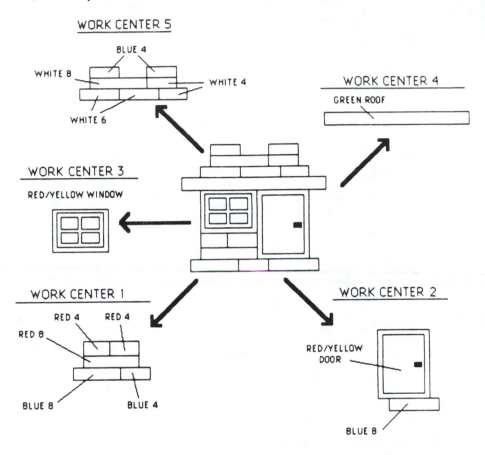

WIP delays. Workers can also help each other during times of idleness to reduce bottlenecks in the production cell. Worker 1 can help Worker 2 or 3 if time permits. Workers should also look for ways of improving production flow and the quality of the work they perform. These ideas should be brought to the attention of the Line Manager by speaking out or waving hands, like an *andon* system.

   2. Material Handlers There are five different material handlers, one each for the five work centers that make up the JIT operation layout in Exhibit 1. Each material handler will be asked to disassemble or break down a set of component parts of units of either Product A or Product B. The specific component part disassembly work for each work center is presented in Exhibit 2 for Product "A" and Exhibit 3 for Product "B." For example, for Product "A," Material Handler 5 for Work Center 5 has to disassemble only the seven bricks on the top of the toy

EXHIBIT 3 • *LEGO™ Brand Toy House Product "B" Instructions for JIT Cell Layout*

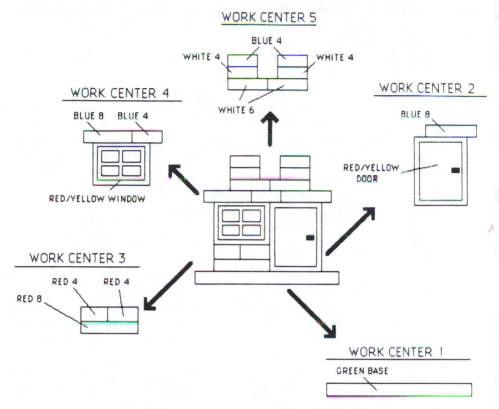

house finshed product. The material handlers simply undo the work performed by their respective workers. To facilitate visual production instructions, students may keep the exhibits next to their open book for reference. They must be sure to disassemble all of the individual bricks and not leave any of the bricks together. They should place the bricks individually in the trays provided for each of them as quickly as possible. This will simulate JIT unitary arrivals of inventory. If problems occur, they should be brought to the attention of the Line Manager as soon as possible by speaking out or waving hands, like an *andon* system.

3. Line Manager The Line Manager is responsible for managing the entire operation, except the Quality Control Inspector, the Inventory Accountant, and the Time Study Engineer. The Line Manager reports only to the President. In the beginning of each simulation game the Line Manager must ensure all workers are trained to do their jobs. Consistent with JIT principles, during the game, the manager should look for problems to be solved in production flow, and deal with any problems workers may bring or observe. The Line Manager's decision is final and must be obeyed by all workers and material handlers. If problems occur, the manager is free to stop a production cell to deal with them, or if possible, implement changes while production continues. The Line Manager gives defective items found by the Quality Control Inspector directly back to the worker whose fault it is, for correction.

4. Quality Control Inspector The Quality Control Inspector is responsible for making sure that each finished product is consistent with the requirements in their respective exhibit. While not always required in JIT systems, the added inspection serves to simulate customer-found defects. If defects are found, the work center where the defect originates is to be given the finished unit by the Quality Control Inspector for breakdown and correction. The Quality Control Inspector is also required to record the number of defects by work center for the entire simulation. (A piece of paper and pencil are necessary for this recording task.) Acting more like a customer quality control system, the Quality Control Inspector reports directly to the President and is not under the supervision of the Line Manager.

5. Inventory Accountant The Inventory Accountant will be asked to perform a number of specific counting and recording tasks by the President. (A piece of paper and pencil are necessary for this recording task.) The recording tasks will vary by the nature of the simulation game being played. The Inventory Accountant reports directly to the President and is not under the supervision of the Line Manager.

6. Time Study Engineer The Time Study Engineer will be asked to perform a number of specific timing and recording tasks by the President. (A stopwatch, piece of paper, and pencil are necessary for this recording task.) The recording tasks will vary by the nature of the simulation game being played. The Time Study Engineer reports directly to the President and is not under the supervision of the Line Manager.

The Basic EOQ Operation

The basic EOQ operation (hereafter called the EOQ operation) will be used in all of the simulation experiments with little or no modification. The layout of the EOQ operation will consist of a continuous flow shop cell or assembly line layout of ten work centers as presented in Exhibit 4. The EOQ operation will operate under a large-lot inventory system and seek to have the fairly specialized work tasks (few bricks to assemble) characteristic of EOQ systems. Work on either of the two products will begin in Work Center 1 and proceed to Work Center 10 for completion. From there, the finished product will be given to the

Quality Control Inspector for inspection. Once approved, the finished product will be given to Material Handler 10, who will start the breaking down process of the finished product. The component inventory items the material handlers obtained will be temporarily stored until a lot size of 5 units is obtained. When a 5-unit lot size has accumulated, the lot of 5 is given by the material handlers back to their respective work centers. This helps to simulate the lot size ordering charactertistic of EOQ operations. This completes the process of production and inventory replenishment, which will permit continuous activity from all participants. The basic roles of each of the six different types of players in the game are as follows:

1. Workers   There are ten different workers, one each for the ten work centers that make up the EOQ operation layout in Exhibit 4. Each worker will be asked to assemble a set of component parts into units of either Product A or Product B. The specific component part assembly work for each work center is presented in Exhibit 5 for Product A and Exhibit 6 for Product B. For example, for Product A the worker at Work Center 1 has to assemble a door and one LEGO™ brick. The color of the brick and the size, represented by the number of locking nodes on top of each brick, are indicated in the exhibits. Even though it is assumed that students will have read these instructions, workers are not to use them in the performance of their jobs. Instead, students performing the role of workers should follow only those instructions given by their Line Manager. Workers are also asked not to perform any quality control activities, because these are not characteristically part of the highly specialized assembly line operation of an EOQ operation. Trying to assemble the right color and size brick in the right place is what counts for quality. If defects in WIP are found by a worker they are to be ignored, since they are the Quality Control Inspector's and Line Manager's concern, not the worker's. If directed to correct defects by the Line Manager, workers will have to break down finished products the same way they put them together to reach and correct defects. This means the defective products will be sent back down the assembly line from worker to worker who must then undo their contributions until it reaches the work center where the defect is corrected. Once the defect is corrected on the line, that item is sent back through the line, just like any other WIP. When the game begins, a total supply of inventory will be sitting in as many trays as the worker has component parts to assemble. For example, if a worker has only two parts to assemble, there will be two trays for inventory. Each component part should have its own individual tray and should never be mixed together with different component parts. The Line Manager will position inventory trays, and neither the workers nor the material handlers should move the trays to facilitate jobs. Plant layout is up to the Line Manager and no one else. Workers are asked not to reorganize them or perform any preassembly work on inventory. Don't preassemble any parts, just leave them as individual inventory items. Ideas for improving the layout efficiency will be up to Line Managers.

EXHIBIT 4   •   *EOQ Production Cell Layout and Staff*

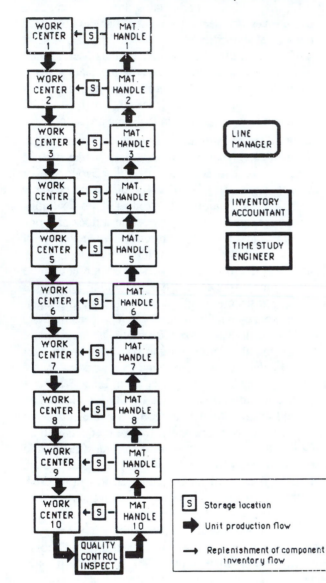

LINE MANAGER

INVENTORY ACCOUNTANT

TIME STUDY ENGINEER

| S | Storage location |

➡ Unit production flow

→ Replenishment of component inventory flow

2. Material Handlers There are ten different material handlers, one each for the ten work centers that make up the EOQ operation layout in Exhibit 4. Each material handler will be asked to disassemble or break down a set of component parts of units of either Product A or Product B in the same way as the JIT material handlers operate. The specific component part disassembly work for each work center is presented in Exhibit 5 for Product A and Exhibit 6 for Product B. For example, for Product A, Material Handler 4 for Work Center 4 has to disassemble only three bricks on the toy house finished product. The material handlers simply undo the work that their respective worker did previously. While the material handlers are assumed to have read these instructions, they should not use them for reference once the game begins. The material handlers will be instructed on their job activities and managed by the Line Manager. Be sure to disassemble all of the individual bricks. Don't leave any of the bricks together. Place the bricks individually in a location where workers can't reach them until a lot size of 5 pieces has been reached. Once reached, the lot of 5 bricks, doors, or windows should be placed in the worker trays provided for each of them. This will simulate the EOQ lot size arrivals of inventory.

EXHIBIT 5 • *LEGO™ Brand Toy House Product "A" Instructions for EOQ Cell Layout*

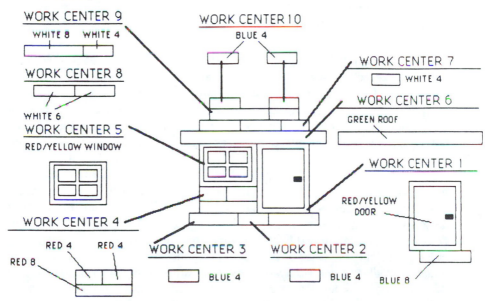

EXHIBIT 6 • *LEGO™ Brand Toy House Product "B" Instructions for EOQ Cell Layout*

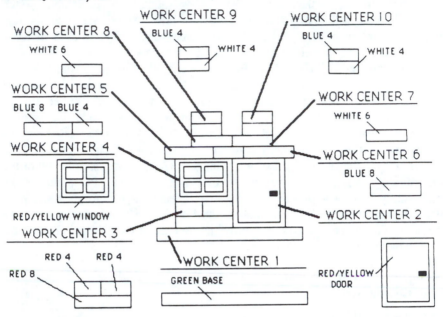

3. Line Manager The Line Manager is responsible for managing the entire operation except the Quality Control Inspector, the Inventory Accountant, and the Time Study Engineer. The Line Manager reports only to the President. Plant layout design is up to Line Managers, no one else. The Line Manager will position all inventory trays and neither the workers nor the material handlers should move the trays. In the beginning of each simulation game the Line Manager must ensure everybody is trained to do their jobs. Ideas for improving the layout efficiency, production flow, or inventory will be solely up to the Line Manager. During the game, the manager is responsible for production flow and dealing with any worker problems observed. Remember, workers in an EOQ system are paid to do a specific job, not to identify or solve the Line Manager's problems. If defects in finished products are found by the Quality Control Inspector, the Line Manager must rechannel the finished product back down the line of workers (not down the material handlers' line). Each worker will break down the finished part until the defective brick is located and corrected. Even though this quality control correction activity will disrupt the flow of the line, it helps simulate rework activities in EOQ operations.

4. Quality Control Inspector The Quality Control Inspector is responsible for making sure that each finished product is consistent with the requirements in the respective exhibits. If defects are found, the defective item is given to the Line Manager for correction. The Quality Control Inspector should tell the Line Manager where the defect is on the item. The defective product should then, under the direction of the Line Manager, begin a reverse course down the assembly line for breakdown and correction. The Quality Control Inspector is also required to record the number of defects by work center for the entire simulation. (A piece of paper and pencil are necessary for this recording task.) The Quality Control Inspector reports directly to the President and is not under the supervision of the Line Manager. While this is not always the way it is in real EOQ operations, this will ensure an accurate comparison for the simulation experiment (and prevent the EOQ operation from cheating on product quality standards in favor of production quotas, which does happen in the real world).

5. Inventory Accountant The Inventory Accountant will be asked to perform a number of specific counting and recording tasks by the President. (A piece of paper and pencil are necessary for this recording task.) The recording tasks will vary by the nature of the simulation game being played. The Inventory Accountant reports directly to the President and is not under the supervision of the Line Manager.

6. Time Study Engineer The Time Study Engineer will be asked to perform a number of specific timing and recording tasks by the President. (A stopwatch, a piece of paper, and a pencil are necessary for this recording task.) The recording tasks will vary by the nature of the simulation game being played. The Time Study Engineer reports directly to the President and is not under the supervision of the Line Manager.

Experiments in JIT and EOQ Operations

The purpose of each of the following experiments will be made clear by the President after the simulation is completed. In each experiment both the JIT operation and the EOQ operation will be used to produce or simulate the production of toy products. After both production systems have been simulated, the observed and recorded results from the players will be discussed. The experiments are to be viewed as independent from one another, and changes from the basic JIT and EOQ operations required in one experiment are not to be carried over to the next experiment, unless otherwise directed to do so.

Experiment 1

Set up the basic JIT and EOQ operation layouts. Each operation is to simulate the manufacture of 50 Product A toy houses, then stop.

Experiment 2

Remove the equivalent inventory of 5 finished products from the total supply of 20 units (permitting a total inventory of only 15 units to exist in the system). With this adjustment to total inventory supplies, set up the basic JIT and EOQ operation layouts. Each operation is to simulate the manufacture of 50 Product A toy houses, then stop.

Experiment 3

Remove the equivalent inventory of 10 finished products from the total supply of 20 units (permitting a total inventory of only 10 units to exist in the system). With this adjustment to total inventory supplies, set up the basic JIT and EOQ operation layouts. Each operation is to simulate the manufacture of 50 Product A toy houses, then stop.

Experiment 4

With only one exception, set up the basic JIT and EOQ operation layouts, and manufacture 50 Product A toy houses, then stop. The exception is that the material handlers' jobs for this experiment are to be changed to permit them to preassemble component parts into subassemblies for the workers' use in production. When the game begins, all units of inventory should be preassembled into subassemblies. For example, Worker Center 1 should have 20 units of subassemblies consisting of five assembled bricks, as presented in Exhibit 2, available for inventory. This helps to simulate the possible pre-production activities that can be a negotiated part of vendor contracted services to aid production flow.

Experiment 5

Repeat Experiment 4, but remove the equivalent inventory of 5 finished products from the total supply of 20 units (permitting a total inventory of only 15 units to exist in the system).

Experiment 6

Repeat Experiment 4 but remove the equivalent inventory of 10 finished products from the total supply of 20 units (permitting a total inventory of only 10 units to exist in the system).

Experiment 7

Set up the basic JIT operation layout, and manufacture 50 Product A toy houses, then stop. Reorganize work activities and inventory to set up the basic JIT operation layout, and manufacture 50 Product B toy houses, then stop. That finishes the JIT simulation efforts. Set up the basic EOQ operation layout, and manufacture 50 Product A toy houses, then stop. Reorganize work activities and inventory to set up the basic EOQ operation layout, and manufacture 50 Product B toy houses, then stop. A total of 2 lots of 50 units each of the two products (100 units of each of the products or 200 units of products in all) will be produced.

Experiment 8

Set up the basic JIT operation layout, and manufacture 25 Product A toy houses, then stop. Reorganize work activities and inventory to set up the basic JIT operation layout, and manufacture 25 Product B toy houses, then stop. Reorganize work activities and inventory to set up the basic JIT operation layout and manufacture 25 Product A toy houses, then stop. Reorganize work activities and inventory to set up the basic JIT operation layout, and manufacture 25 Product B toy houses, then stop. That finishes the JIT simulation efforts. Set up the basic EOQ operation layout, and manufacture 25 Product A toy houses, then stop. Reorganize work activities and inventory to set up the basic EOQ operation layout, and manufacture 25 Product B toy houses, then stop. Reorganize work activities and inventory to set up the basic EOQ operation layout, and manufacture 25 Product A toy houses, then stop. Reorganize work activities and inventory to set up the basic EOQ operation layout, and manufacture 25 Product B toy houses, then stop. A total of 4 lots of 25 units each of the two products (100 units of each product or 200 units of the products in all) will be produced.

Experiment 9

Repeat Experiment 8, except this time produce a total of 5 lots of 10 units each of the two products with the intermixed scheduling as before between the two products. A total of 5 lots of 10 units each of the two products (50 units of each product or 100 units of the products in all) will be produced.

Experiment 10

Repeat Experiment 8, but now remove the equivalent inventory of 5 finished products from the total supply of 20 units (permitting a total inventory of only 15 units to exist in the system).

Experiment 11

Repeat Experiment 8, but now remove the equivalent inventory of 10 finished products from the total supply of 20 units (permitting a total inventory of only 10 units to exist in the system).

Experiment 12

Repeat Experiment 11 with the EOQ operation performing the work first. Direct the EOQ manager to concentrate on developing ideas to improve setup time and production activities. None of these ideas is to cause the layout to have less than 10 work centers. This allows the simulated specialization of work tasks typically required in this type of operation. If no suggestions are offered, the EOQ operation can stop. If there are suggestions, implement them and repeat Experiment 11 again. Direct JIT workers and the JIT line manager to concentrate on the encouragement of developing ideas to improve setup time and production activities. None of these ideas is to cause the layout to have more than five work centers. This allows the simulation of workers with enlarged job tasks characteristic of a JIT operation. If no suggestions are offered, the JIT operation can stop. If there are suggestions, implement them and repeat Experiment 11 again.

Index